T0213890

Communications in Computer and Information Science 1092

Commenced Publication in 2007
Founding and Former Series Editors:
Phoebe Chen, Alfredo Cuzzocrea, Xiaoyong Du, Orhun Kara, Ting Liu,
Krishna M. Sivalingam, Dominik Ślęzak, Takashi Washio, Xiaokang Yang,
and Junsong Yuan

Editorial Board Members

Simone Diniz Junqueira Barbosa ⓘ
  Pontifical Catholic University of Rio de Janeiro (PUC-Rio),
  Rio de Janeiro, Brazil
Joaquim Filipe ⓘ
  Polytechnic Institute of Setúbal, Setúbal, Portugal
Ashish Ghosh
  Indian Statistical Institute, Kolkata, India
Igor Kotenko ⓘ
  St. Petersburg Institute for Informatics and Automation of the Russian
  Academy of Sciences, St. Petersburg, Russia
Lizhu Zhou
  Tsinghua University, Beijing, China

More information about this series at http://www.springer.com/series/7899

Aleš Zamuda · Swagatam Das ·
Ponnuthurai Nagaratnam Suganthan ·
Bijaya Ketan Panigrahi (Eds.)

Swarm, Evolutionary, and Memetic Computing and Fuzzy and Neural Computing

7th International Conference, SEMCCO 2019
and 5th International Conference, FANCCO 2019
Maribor, Slovenia, July 10–12, 2019
Revised Selected Papers

 Springer

Editors
Aleš Zamuda 🆔
University of Maribor
Maribor, Slovenia

Ponnuthurai Nagaratnam Suganthan 🆔
Division of Control and Instrumentation
Nanyang Technological University
Singapore, Singapore

Swagatam Das 🆔
Electronics and Communication
Sciences Unit
Indian Statistical Institute
Kolkata, West Bengal, India

Bijaya Ketan Panigrahi 🆔
Department of Electrical Engineering
Indian Institute of Technology Delhi
New Delhi, Delhi, India

ISSN 1865-0929   ISSN 1865-0937 (electronic)
Communications in Computer and Information Science
ISBN 978-3-030-37837-0   ISBN 978-3-030-37838-7 (eBook)
https://doi.org/10.1007/978-3-030-37838-7

This Springer imprint is published by the registered company Springer Nature Switzerland AG
The registered company address is: Gewerbestrasse 11, 6330 Cham, Switzerland

Preface

This volume contains the papers presented at SEMCCO 2019 and FANCCO 2019: Swarm, Evolutionary and Memetic Computing Conference (SEMCCO 2019) and Fuzzy And Neural Computing Conference (FANCCO 2019), held during July 10–12, 2019, in Maribor, Slovenia. SEMCCO 2019 was the 7th international conference of this series, where SEMCCO 2010 was successfully organized at SRM University, Chennai, SEMCCO 2011 at ANITS, Visakhpatnam, SEMCCO 2012 at SOA University, Bhubaneswar, SEMCCO 2013 at SRM University, Chennai, SEMCCO 2014 at SOA University, Bhubaneswar, and SEMCCO 2015 in Hyderabad, India. FANCCO 2019 was the 5th international conference of this series, collocated and co-organized with SEMCCO. These conferences aim at bringing together researchers from academia and industry to report and review the latest progresses in cutting-edge research, focusing on Swarm, Evolutionary, Memetic, Fuzzy, and Neural computing to explore new application areas and to design new bio-inspired algorithms for solving specific, hard optimization problems, and finally to create awareness of these domains to a wider audience of practitioners. Therefore, researchers are encouraged to submit their contributions in both theoretical and practical aspects. This year, the General Chair of the conference was Aleš Zamuda, the General Co-chairs were Swagatam Das (also Program Chair), Ponnuthurai Nagaratnam Suganthan (also Steering Committe Chair), and Bijaya Ketan Panigrahi (also Publication Chair).

There were 31 submissions. Each submission underwent a single-blind review by at least 3, and on the average 3.0, Program Committee members. The committee decided to accept 18 papers for this post-conference proceedings. The conference program also included an invited talk by Ponnuthurai Nagaratnam Suganthan from Nanyang Technological University on "Differential Evolution for Numerical Optimization," an invited talk by Benjamin Doerr from the Max-Planck-Institut für Informatik on "From Theory to Better Algorithms," and a tutorial by Aleš Zamuda from the University of Maribor on "Differential Evolution Applicability." The conference included 5 sessions with paper presentations, a benchmarking panel discussion chaired by Aleš Zamuda, and was combined with 4 social events. The conference also included competitions on numerical optimization with entries for the 100-Digit Challenge.

We take this opportunity to thank the authors of all submitted papers for their hard work, adherence to the deadlines, and patience with the review and publishing process. As the quality of a refereed volume depends largely on the expertise and dedication of the reviewers, we thank the Program Commitee members who produced excellent reviews. We would also like to thank our sponsors and acknowledge Springer, the University of Maribor, IEEE Slovenia, COST, and EasyChair for their support of this conference. We thank Management and Administrations (faculty colleagues and administrative personnel) of the University of Maribor at the Faculty of Electrical Engineering and Computer Science. We would also like to thank the participants of this conference. Finally, we would like to thank all the volunteers for their tireless efforts in meeting the deadlines and arranging every detail to make sure that the conference ran smoothly. We hope that the readers of these proceedings find the papers inspiring and enjoyable.

November 2019

Aleš Zamuda

Swagatam Das

Ponnuthurai Nagaratnam Suganthan

Bijaya Ketan Panigrahi

Organization

Program Committee

Ying-Ping Chen	National Chiao Tung University, Taiwan
Swagatam Das	Indian Statistical Institute, India
Kusum Deep	Indian Institute of Technology Roorkee, India
Tome Eftimov	Stanford University, USA
Mohammed El-Abd	American University of Kuwait, Kuwait
Steffen Finck	Vorarlberg University of Applied Sciences, Austria
Heiko Hamann	University of Lübeck, Germany
Daniel Hernandez	ULPGC, Spain
Mauro Iacono	Università degli Studi della Campania Luigi Vanvitelli, Italy
Zuzana Kominkova Oplatkova	Tomas Bata University in Zlin, Czech Republic
Antonio Latorre	Universidad Politécnica de Madrid, Spain
Simone Ludwig	North Dakota State University, USA
Karol Opara	Systems Research Institute, Polish Academy of Sciences, Poland
Bijaya Ketan Panigrahi	IIT Delhi, India
Adam Piotrowski	Polish Academy of Sciences, Poland
S. G. Ponnambalam	University Malaysia Pahang, Malaysia
Mallipeddi Rammohan	Kyungpook National University, South Korea
Ponnuthurai Nagaratnam Suganthan	Nanyang Technological University, Singapore
Ryoji Tanabe	Southern University of Science and Technology, China
Fatih Tasgetiren	Yasar University, Turkey
Ankit Thakkar	Institute of Technology at Nirma University, India
Daniela Zaharie	West University of Timisoara, Romania
Aleš Zamuda	University of Maribor, Slovenia
Roman Šenkeřík	Tomas Bata University in Zlin, Czech Republic

Additional Reviewers

Alić, Amina
Biswas, Partha
Brest, Janez
Bujok, Petr
Burguillo, Juan Carlos
Campanile, Lelio
Ding, Weiping
Fister, Dusan
Gao, Kaizhou
Ghosh, Arka

Gupta, Avisek
Macků, Lubomír
Mastroianni, Michele
Mullick, Sankha Subhra
Pluhacek, Michal
Stankovski, Vlado
Tanveer, Mohammad
Viktorin, Adam
Wang, Yong

Contents

Cooperative Model of Evolutionary Algorithms and Real-World Problems

Petr Bujok$^{(\boxtimes)}$ (iD)

University of Ostrava, 30. dubna 22, 70200 Ostrava, Czech Republic
petr.bujok@osu.cz

Abstract. A cooperative model of efficient evolutionary algorithms is proposed and studied when solving 22 real-world problems of the CEC 2011 benchmark suite. Four adaptive algorithms are chosen for this model, namely the Covariance Matrix Adaptation Evolutionary Strategy (CMA-ES) and three variants of adaptive Differential Evolution (CoBiDE, jSO, and IDEbd). Five different combinations of cooperating algorithms are tested to obtain the best results. Although the two algorithms use constant population size, the proposed model employs an efficient linear population-size reduction mechanism. The best performing Cooperative Model of Evolutionary Algorithms (CMEAL) employs two EAs, and it outperforms the original algorithms in 10 out of 22 real-world problems.

Keywords: Evolutionary algorithm · CoBiDE · jSO · IDEbd · CMA-ES · Cooperation · Real-world problems

1 Introduction

In this paper, a cooperative model of Evolutionary Algorithms (EA) is proposed and studied. The main idea is to employ different evolutionary approaches of population development to achieve better results. Although EAs provide an acceptable solution in reasonable computational time due to the stochastic character, they cannot guarantee an acceptable solution in a finite computational time.

One of the most widely used EA in the last decades is admittedly Differential Evolution (DE) algorithm [14]. The main reason for the high popularity of the DE algorithm is in its simplicity and efficiency. Therefore, a lot of variants are the leading optimisers and DE has been developed very intensively [8,10,13].

Although DE is very efficient, there is still no optimisation algorithm which is the most efficient for all the global optimisation problems (No Free Lunch theorem [18]). One of the well-performing 'non-DE' algorithms is the Evolution Strategy with Covariance matrix adaptation approach (CMA-ES) [12]. In 2016, Bischl et al. proposed an approach to select a proper optimisation method to solve different tasks [1]. Here, a cooperation of several EAs is studied.

The rest of the paper is organised in the following manner. The newly proposed cooperative model of Evolutionary Algorithms is presented in Sect. 2.

© Springer Nature Switzerland AG 2020
A. Zamuda et al. (Eds.): SEMCCO 2019/FANCCO 2019, CCIS 1092, pp. 1–12, 2020.
https://doi.org/10.1007/978-3-030-37838-7_1

Experimental settings and the achieved results are represented in Sects. 3 and 4. The paper is briefly concluded in Sect. 5.

2   A Cooperative Model of Evolutionary Algorithms

In a latter study [7], newly proposed model employing four various well-known optimisation algorithms – CoBiDE [17], IDEbd [3], CMA-ES [12], and jSO [2] was introduced. The main goal is to use the most successful EA to develop the population in the current moment of the search process. In other words, the method providing better results is more prioritised in the next generation. The success of employed EAs is based on a number of new-good individuals in the preceding stages. When a presently used EA has to be displaced using more successful one, the development of the population is carried out by a new EA. The mechanism of competition of employed EAs was introduced in 2006 by Tvrdík, where several various DE parameters settings are used to be selected in the development of the population [16].

Other models of cooperation of various optimisation algorithms were studied in our recent research. The cooperative model of six different state-of-the-art DE algorithms was proposed in 2012 [6]. The cooperation of the employed DE variants was studied more deeply when it solves 22 real-world problems CEC 2011 [5]. The model using cooperation of eight various nature-inspired algorithms was applied to 22 real-world problems [4].

Although the proposed cooperative model uses a mechanism for the competition of four EAs, several combinations of employed algorithms are applied. A more proper subset of algorithms will be selected studying the real successes of the employed techniques. Because, two out of four employed algorithms (jSO and IDEbd) are used in the final model, these methods are discussed in more detail. The two remaining employed algorithms are described briefly, more details are provided in the original papers.

2.1   CoBiDE

In 2014, a new DE variant with 'covariance-matrix learning' called CoBiDE was proposed [17]. CoBiDE adds two new aspects to a classic DE. The crossover in CoBiDE employes a covariance-matrix learning approach. This approach is used for rotation of the coordinate system. The rotation controls an adaptation of the dependencies in the population. In addition, it promises higher efficiency on the rotated objective functions. Similarly, the CMA-ES algorithm uses a rotation-invariant approach, and it is employed in the studied cooperative model.

The second element newly used in CoBiDE is the bimodal sampling of F_i and CR_i parameters, which distinguishes between exploration and exploitation. The initial values of F_i and CR_i are set for each ith individual using the Cauchy distribution:

$$F_i = \begin{cases} \text{randc}(0.65, \ 0.1) & \text{if rand}(0, \ 1) < 0.5 \\ \text{randc}(1.0, \ 0.1) & \text{otherwise,} \end{cases} \tag{1}$$

$$CR_i = \begin{cases} \text{randc}(0.1, \ 0.1) & \text{if rand}(0, \ 1) < 0.5 \\ \text{randc}(0.95, \ 0.1) & \text{otherwise.} \end{cases} \tag{2}$$

CoBiDE uses the most popular original rand/1 mutation [14]. After the mutation, the covariance-matrix based crossover is applied to the whole population with probability controlled by the parameter pb. In other cases, the standard binomial crossover is used to whole population in the current generation. The values of F_i and CR_i are re-sampled if the new ith trial solution is not successful. Before each generation, Eigenvalues (D) and Eigenvectors (B) are extracted from the covariance matrix (C) of current population.

$$C = BD^2B^T. \tag{3}$$

Then, a trial point y_i' is computed in a new 'eigenvector' coordinate system employing the transformed parent and mutation vectors:

$$x_i' = B^{-1}x_i = B^Tx_i, \tag{4}$$

$$u_i' = B^{-1}u_i = B^Tu_i. \tag{5}$$

Then, a binomial crossover is used to develop a trial solution in a new coordinate system. A new individual y_i' is transformed back into a standard coordinate system:

$$y_i = By_i'. \tag{6}$$

This method achieves better results for problems where the population is strongly correlated. The authors of CoBiDE recommended to employ this approach occasionally, controlled by the parameter pb. The Eigenvectors and Eigenvalues are extracted only from a part of individuals with a lower function value of the population. The portion of the selected individuals of the population is controlled by the second parameter ps.

2.2   Individual-Dependent Approach

The second DE algorithm employed in the studied cooperative model is an enhanced variant of the original IDE algorithm (IDEbd) proposed in [3]. Despite the original IDEbd applies diversity-based population size control, the cooperative model employs IDEbd without this mechanism. The search process of the original IDEbd algorithm [15] is divided into two stages – more explorative and more exploitative. A parameter F influences the size of the searching area of the base individual x_o (Eq. (9)). The population of IDEbd is sorted with respect to the objective function values. The individuals are distributed into two sets - superior S (smaller objective function) and inferior I (higher objective function). The values of the parameters F and CR are sampled by (7) and (8):

$$F_o = \frac{o}{N}, \tag{7}$$

$$CR_i = \frac{i}{N},\tag{8}$$

where o is an index of a base point of mutation, i represents an index of the current point, and N denotes the population size. The index of the base vector depends on the stage, $o = i$ in the first stage of the search, and o is selected randomly in the second stage. Values of F and CR are modified by normal distribution with variance 0.1, until they are in the interval $(0, 1)$.

The IDEbd variant enhances the original IDE mutation scheme as shown in Eq. (9). The first part of the mutation uses a base individual $(\boldsymbol{x}_o)$ as the current point of the population at the exploration phase, and the base vector selected randomly is used in the second exploitation stage. It can increase the diversity of the population in the early stage and accelerate convergence in the last phase.

$$\boldsymbol{u}_i = \begin{cases} \boldsymbol{x}_o + F_o * (\boldsymbol{x}_{r_1} - \boldsymbol{x}_o) + F_o * (\boldsymbol{x}_{r_2} - \boldsymbol{x}_{r_3}) & \text{if } o \in S \\[2mm] \boldsymbol{x}_o + F_o * (\boldsymbol{x}_{\text{better}} - \boldsymbol{x}_o) + F_o * (\boldsymbol{x}_{r_2} - \boldsymbol{x}_{r_3}) & \text{if } o \in I, \end{cases}\tag{9}$$

where o is denotes the base individual, $r_1 \neq r_2 \neq r_3 \neq o$ are indices selected randomly from $[1, N]$, and $better$ represents a randomly selected index from the superior portion of the individuals S.

The value of ps controls the part of superior population, and it is adapted using:

$$ps = 0.1 + 0.9 \times 10^{5 \times \left(\frac{g}{g_{\max}} - 1\right)},\tag{10}$$

where g is the current generation, and $g_{\max}$ is the maximum number of generations.

The last individual in the mutation, $\boldsymbol{x}_{r_3}$, is further perturbed with small probability p_d. It enables the base vector extricate from the local area.

$$x_{r_3,j} = \begin{cases} a_j + rand(0,1) \times (b_j - a_j), & \text{if } rand_j(0,1) < p_d \\[2mm] x_{r_3,j} & \text{otherwise}, \end{cases}\tag{11}$$

where $p_d = 0.1 \times ps$ and a_j, b_j are the boundaries of the jth coordinate. Then, standard binomial crossover is applied to generate new individuals.

2.3   CMA-ES

The Evolutionary Strategy with Covariance Matrix Adaptation (CMA-ES) algorithm used in the studied model was proposed by Hansen and Ostermeier in [12]. A new trial point $\boldsymbol{x}^N$ is generated using only a mutation strategy. It adds a random vector to the current point $\boldsymbol{x}^E$:

$$\boldsymbol{x}^N = \boldsymbol{x}^E + \sigma \boldsymbol{B} N(\boldsymbol{0}, \boldsymbol{I}).\tag{12}$$

A part $\boldsymbol{B} N(\boldsymbol{0}, \boldsymbol{I})$ represents a linear transformation of $N(\boldsymbol{0}, \boldsymbol{I})$ for arbitrary matrix B with size of $D \times D$. Choosing B in a propoer way, arbitrary normal distribution with a zero mean vector can be generated by the transformation (12).

A parameter σ controls the variance in the mutation. More details of the well-performing CMA-ES algorithm are provided in [12].

The population in CMA-ES is developed differently than in DE. Only a weighted 'centre' of the present population is used as a seed in the next generation. Therefore, the whole population in a new generation is generated from the same base position. However, each individual is generated using the mutation (12). Such a difference between DE and CMA-ES approaches originates a special condition when EAs are changed to develop the population.

2.4 jSO

An efficient DE variant derived from the successful JADE, SHADE, L-SHADE, and iL-SHADE is called jSO [2]. The jDE algorithm was introduced in 2017 at the CEC 2017 competition when it took the second position. jSO uses historical circle memories of the control parameters of length 5, initialised by the mean values $\mu_F = 0.3$, and $\mu_{CR} = 0.8$ (the last settings are replaced with the values $F = 0.9$ and $CR = 0.9$). The authors of jSO proposed to prioritise higher values of CR in the first half of the search and keep values of F under 0.7 in the first 60% of evaluations. These restrictions make jSO very efficient; nevertheless, there is no more place to improve this method.

The jSO algorithm uses an updated current-to-$p$best mutation strategy, which is controlled by a weighted F parameter:

$$u_i = x_i + F_w(x_{\text{pBest}} - x_i) - F(x_{r1} - x_{r2}), \tag{13}$$

where

$$F_w = \begin{cases} 0.7F, & FES < 0.2maxFES \\ 0.8F, & FES < 0.4maxFES \\ 1.2F, & \text{otherwise,} \end{cases} \tag{14}$$

where FES is the current number of function evaluations, and $maxFES$ denotes the maximal allowed number of function evaluations. Individuals for mutation are selected as follows – x_i is the current point, x_{pBest} is selected randomly from $p \times 100\%$ best points of P, and x_{r1} and x_{r2} are randomly selected points from P and $P \cap A$, respectively. An archive A of the size $N \times 2.6$ is used to store good old solutions.

Then, a standard binomial crossover is used. The values of F and CR are computed from the recent successful settings using a weighted mean. A linear reduction of the population size is used with the initial population size $N = 25 \times \sqrt{D} \times \log D$. The parameter p is linearly decreased from 0.25 to 0.125. More details about the jSO algorithm are provided in the original paper [2].

2.5 Competitive Mechanism and Population-Size Reduction

After initialisation and evaluation of the population, the control parameters of involved EAs are initialised. Besides this, probabilities to use of each hth EA are set equally to $q_h = 1/H$, and H denotes the number of employed EAs.

A selection of the algorithm to evaluate the population is performed randomly, using the probabilities q_h. Simply, an algorithm with higher probability q_h has a higher chance to be selected in the next generation, and vice versa. If a currently used algorithm generates successful individual ($f(\boldsymbol{y}_i) \leq (\boldsymbol{x}_i)$), the count of new successful individuals is increased by one. At the end of generation, the probabilities q_h are updated for the next generation:

$$q_h = \frac{n_h + n_0}{\sum_{j=1}^{H}(N_j + N_0)},\tag{15}$$

where n_h is the current count of the hth EA successes. The input parameter $n_0 = 2$ is used against a dramatic primacy of q_h by one random successful use of the hth parameter setting. The current values of q_h are initialised $q_h = 1/H$ if any value q_h decreases below the input limit $\delta = 1/(5 * H)$.

Beside the competition mechanism, the population size mechanism from jSO is also used in the proposed model. At the end of generation, the proper population size is updated, and the population size is reduced:

$$N = round[(\frac{N_{\min} - N_{\text{init}}}{maxFES})FES + N_{\text{init}}],\tag{16}$$

where FES is the present number of function evaluations, N_{init} represents the initial population size, $N_{\min}$ denotes the population size at the end of run when the allowed number of $maxFES$ function evaluations is achieved.

The studied cooperative model of Evolutionary Algorithms with a linear population size reduction is denoted $CMEAL$, in the presentation of results abbreviated variant CM is used.

3   Experimental Settings

In this paper, several cooperative CMEAL variants are studied, based on successfully generated individuals. At first, two models using all four EAs are proposed with a different population size at the end of the search, $N_{\min} = 5$, 20, and they are called $CM4_{\text{N5}}$ and $CM4_{\text{N20}}$. Next model uses only three EAs (CMA-ES, IDEbd, jSO) and the value of $N_{\min}$ is set according to $D - N_{\min} = 5$ if $D < 20$; $N_{\min} = 10$ if $20 \leq D < 40$; $N_{\min} = 20$ if $D \geq 40$ ($CM3_D$). Further model ($CM3_{\text{DE}}$) uses only three DE variants (CoBiDE, IDEbd, jSO) and the same settings of $N_{\min}$ based on D. The last proposed model uses only two most successful DE versions IDEbd and jSO, and the same setting of $N_{\min}$ – $CM2_D$. The only control parameter of CMEAL is a number of employed EAs, $H = 2$, 3, 4. Then, $\delta = 1/(5 * H) = 0.1$, 0.066, 0.05.

The experimental comparison is based on a test suite of 22 real-world tasks from the CEC 2011 competition in the Special Session on Real-Parameter Numerical optimisation [9]. The computational complexity of the problems is various, as soon as the dimensionality of the search space ($D \in (1, 240)$). The experiments are for each algorithm and problem repeated in 25 independent runs. The algorithm stops when it achieves the given number of function evaluation, $MaxFES = 150000$. For more complex analysis, the results of the algorithms at $MaxFES= 50000$ and $MaxFES= 100000$ were kept. The solution of the algorithm on the problem is represented by the individual from the terminal population with the least function value.

The algorithms use a linear population-size reduction with initial value $N_{init} = \text{round}(25 \times \log(D) \times \sqrt{D})$. For problems with a low dimension level $D < 6$, the initial population size is computed from value $D = 6$. All parameters of the four used Evolutionary Algorithms are set to values recommended by the authors.

The $CMEAL$ algorithm is implemented and experimentally tested in Matlab 2017b. All computations were carried out on a standard PC with Windows 7, Intel(R) Core(TM)i7-4790 CPU 3.6 GHz, 16 GB RAM.

4 Results and Discussion

In this paper, five variants of the proposed CMEAL algorithm are compared with the four original EAs using a set of 22 real-world problems. At first, the Friedman test is used to compare the performance of all nine algorithms. The Friedman test provides the mean ranks of the algorithms using the median values of achieved function values. The mean ranks regarding each algorithm and dimension are in Table 1. The null hypotheses on equal efficiency of all nine algorithms were rejected for all three stages of the run with an achieved significance level $p < 5 \times 10^{-5}$.

Table 1. Results of the comparison of the cooperative model from the Friedman tests.

FES	CM2$_D$	jSO	CM3$_D$	CM4$_{N5}$	CM3$_{DE}$	CM4$_{N20}$	IDEbd	CMAES	CoBiDE
50000	4.7	5.5	4.8	5.7	6.3	6.6	**3.3**	3.5	_4.6_
100000	**4.0**	_4.3_	4.4	5.8	5.8	6.1	4.1	4.4	6.3
150000	**3.1**	3.4	_3.8_	4.4	4.5	5.6	6.2	6.9	7.2

The achieved mean ranks show the variance of performance with increasing computational time (measured by FES). The algorithms in Table 1 are ordered based on their performance in the final stage.

In the first stage, three original algorithms are the best performing. From the second stage ($FES= 100,000$), the best performing algorithm in the comparison is cooperative model of jSO and IDEbd. Promising results are also provided by a cooperative model of jSO, IDEbd, and CMA-ES. Three out of four original EAs are the worst performing algorithms regarding all 22 problems.

The Kruskal-Wallis non-parametric one-way ANOVA test was applied to each test problem separately to see more details. It is clear that the performance of all the algorithms in the experiment significantly differs. Moreover, the Dunn's method was applied for multiple comparison (see Table 2). In the third column,

Table 2. Results of the comparison of new CMEAL variants from the Kruskal-Wallis tests.

Fun	D	p	Best	Worst
T01	6	$*$	All except $\rightarrow$	$CM3_D$
T02	30	$***$	$CM4_{N5}$, $CM3_D$	$CM4_{N20}$
T03	1	$\approx$		
T04	1	$***$	$CM3_D$, $CM3_{DE}$, $CM2_D$	$CM4_{N5}$, $CM4_{N20}$
T05	30	$**$	$CM4_{N5}$, $CM2_D$, $CM3_{DE}$	$CM3_D$, $CM4_{N20}$
T06	30	$***$	$CM4_{N5}$, $CM3_{DE}$, $CM2_D$	$CM4_{N20}$
T07	20	$***$	$CM2_D$, $CM4_{N5}$, $CM3_{DE}$	$CM3_D$, $CM4_{N20}$
T08	7	$\approx$		
T09	126	$**$	$CM3_D$, $CM2_D$, $CM4_{N20}$	$CM3_{DE}$, $CM4_{N5}$
T10	12	$\approx$		
T11.1	120	$*$	jSO, $CM3_{DE}$	$CM3_D$, $CM4_{N5}$, $CM4_{N20}$
T11.2	240	$**$	$CM3_D$, $CM2_D$, $CM3_{DE}$	$CM4_{N5}$, $CM4_{N20}$
T11.3	6	$*$	$CM2_D$, $CM3_{DE}$, $CM4_{N20}$	$CM4_{N5}$, $CM3_D$
T11.4	13	$\approx$		
T11.5	15	$**$	$CM3_D$, $CM2_D$	$CM4_{N5}$, $CM4_{N20}$, $CM3_{DE}$
T11.6	40	$\approx$		
T11.7	140	$***$	$CM2_D$	$CM4_{N20}$, $CM4_{N5}$
T11.8	96	$***$	jSO, $CM3_D$	$CM3_{DE}$, $CM4_{N5}$, $CM4_{N20}$
T11.9	96	$***$	jSO, $CM3_D$	$CM3_{DE}$, $CM4_{N20}$, $CM4_{N5}$
T11.10	96	$***$	jSO, $CM3_D$	$CM4_{N5}$, $CM3_{DE}$, $CM4_{N20}$
T12	26	$**$	$CM3_{DE}$, $CM2_D$, $CM3_D$	$CM4_{N5}$, $CM4_{N20}$
T13	22	$*$	$CM3_{DE}$, $CM4_{N5}$	$CM4_{N20}$, $CM3_D$, $CM2_D$

Table 3. Significant wins and loses of all algorithms from the Kruskal-Wallis tests.

#	$CM2_D$	$CM3_{DE}$	$CM3_D$	$CM4_{N5}$	jSO	$CM4_{N20}$	CoBiDE	CMAES	IDEbd
Wins	10	9	9	5	4	2	0	0	0
Loses	1	5	6	10	0	14	0	0	0

there are the significance values of the Kruskal-Wallis test denoted as follows: '$***$' ($p < 0.001$), '$**$' ($p < 0.01$), '$*$' ($p < 0.05$), and '$\approx$' otherwise. The significantly best performing algorithms are in column 'best', the worst performing algorithms are in column 'worst'. In the 5 out of 22 problems, there is no significant difference between the nine algorithms. In other cases, the CMEAL variant mostly achieves the best results, occasionally shared with adaptive jSO. On the other hand, some of the proposed models are occasionally the worst performing methods. The counts of the best positions and the worst positions of the Dunn test are in Table 3. The algorithms are ordered from better performing to worse performing. It is clear that although the original jSO is never the worst performing, the cooperative model of jSO with IDEbd achieves the best results in 10 out of 22 problems, and it is the worst performing only in one problem.

A comparison of all proposed CMEAL variants with the winner of CEC 2011 competition (GA-MPC [11]) provides the following results. Variant $CM4_{N20}$ performs better in 9 problems and worse in 11 problems; variants $CM4_{N5}$, $CM3_D$, and $CM3_{DE}$ perform better in 10 and worse in 10 problems; $CM2_D$ performs better in 11 and worse in 9 problems.

The fundamental information to the development of CMEAL models is provided in a study of successes of employed EAs. The real successes of employed algorithms in each task are represented in percentage values in Table 4. The distribution of the values is changed with problems. Therefore, the total average successes of the employed CMEAL models are computed in the last rows of these tables. It is clear that in $CM4_{N5}$ and $CM4_{N20}$, the best performing EAs are jSO and IDEbd (34% and 35%, respectively). The CMA-ES algorithm is successful only in 11% of the reproduction process and CoBiDE in $\approx$20%. Variants of CMEAL with three EAs provide very similar results, the best performing EA is jSO (44% and 40%, respectively), and the second most efficient algorithm is IDEbd (41% and 36%, respectively). This evolution of reduction of the employed EAs in the cooperative model results in the last model where the most efficient jSO and IDEbd are employed. The successes of these EAs in $CM2_D$ model are very similar ($\approx$50%).

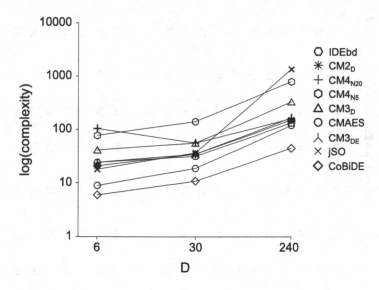

Fig. 1. Estimated time complexity of algorithms in the comparison.

It can be noted that the cooperative model should substantially increase the time complexity. In this paper, the time complexity is estimated by a simulation

Table 4. Frequencies of employed algorithms in CMEAL.

F	D	CM4$_{N5}$				CM4$_{N20}$				CM3$_D$			CM3$_{DE}$			CM2$_D$	
		CoBiDE	IDEbd	CMAES	jSO	CoBiDE	IDEbd	CMAES	jSO	IDEbd	CMAES	jSO	CoBiDE	IDEbd	jSO	IDEbd	jSO
T01	6	30	65	1	4	22	73	1	4	89	2	9	36	61	3	94	6
T02	30	17	33	17	33	19	31	17	33	36	21	44	24	39	37	52	48
T03	1	37	61	1	1	37	62	1	1	44	36	20	40	58	1	62	38
T04	1	38	61	0	1	39	60	0	1	96	1	3	44	55	1	98	2
T05	30	19	29	10	42	20	30	8	42	39	10	51	22	33	45	53	47
T06	30	20	31	4	45	19	33	4	43	41	6	54	19	36	46	50	50
T07	20	21	33	13	33	19	37	13	32	46	16	39	23	41	36	55	45
T08	7	41	56	2	1	28	68	2	2	95	3	2	45	53	2	96	4
T09	126	11	19	21	49	9	11	29	51	22	24	54	18	21	61	19	81
T10	12	15	25	30	30	12	25	26	37	29	24	46	26	39	35	47	53
T11.1	120	8	29	18	46	5	32	17	47	18	13	69	7	32	61	29	71
T11.2	240	9	40	0	50	8	47	0	45	31	0	69	8	24	68	61	39
T11.3	6	36	47	4	13	23	28	15	35	46	26	28	41	47	12	33	67
T11.4	13	25	31	35	9	21	36	33	9	55	35	10	18	13	69	15	85
T11.5	15	18	21	14	46	14	25	12	49	14	12	74	20	29	51	23	77
T11.6	40	12	16	18	54	10	4	21	65	16	17	67	12	26	61	30	70
T11.7	140	15	23	21	41	12	21	19	48	27	30	42	9	44	48	49	51
T11.8	96	11	18	12	59	18	21	11	49	17	17	65	22	22	57	41	59
T11.9	96	12	27	12	48	11	23	12	53	27	11	62	14	32	53	42	58
T11.10	96	9	24	12	56	14	27	12	47	37	13	50	18	23	59	43	57
T12	26	27	29	0	43	27	33	0	39	39	1	60	32	29	39	45	55
T13	22	31	30	2	37	25	34	3	38	41	12	46	25	33	42	49	51
avg.		21	34	11	34	19	35	11	35	41	15	44	24	36	40	49	51

on three problems with various dimensionality ($T01$ with $D = 6$, $T02$ with $D = 30$, and $T11.2$ with $D = 240$). The resulting estimated time complexity in a semi-log scale is depicted in Fig. 1. It is obvious that the most complex method is jSO in the problem $T11.2$ with high dimension. The average complexity from three problems was also computed and methods in the legend of this figure are sorted from the most complex (IDEbd) to the least complex (CoBiDE). The higher time complexity of IDEbd is given by the population-size control mechanism. On the other hand, the small time complexity of CoBiDE is surprising, regarding the Eigenvector crossover mechanism.

5   Conclusion

The newly proposed cooperative model of the four efficient Evolutionary Algorithms provides very good performance. Some CMEAL variants significantly outperform the original EAs in a set of real-world problems. The best performing CMEAL called $CM2_D$ was developed in the gradual study of the four employed EAs, and it uses jSO and IDEbd algorithms. The results show that this model provides very good performance as it achieves the best results in 10 out of 22 problems, and it is the worst performing only in one problem. The best performing jSO is the best performing only in four problems. Besides the efficiency of the mentioned model, there are variants of CMEAL that provide worse results ($CM4_{N20}$ or $CM4_{N5}$). Studying the estimated time complexity of the proposed models, there is no significant increase in computational demands. This information is crucial for further research in this area. The proposed CMEAL model will be studied in more detail, and another EAs will be employed in future work.

References

1. Bischl, B., et al.: ASlib: a benchmark library for algorithm selection. Artif. Intell. **237**, 41–58 (2016)
2. Brest, J., Maučec, M.S., Bošković, B.: Single objective real-parameter optimization: algorithm jSO. In: 2017 IEEE Congress on Evolutionary Computation (CEC), pp. 1311–1318 (2017)
3. Bujok, P., Tvrdík, J.: Enhanced individual-dependent differential evolution with population size adaptation. In: 2017 IEEE Congress on Evolutionary Computation (CEC), pp. 1358–1365, June 2017
4. Bujok, P.: Cooperative model for nature-inspired algorithms in solving real-world optimization problems. In: Korošec, P., Melab, N., Talbi, E.-G. (eds.) BIOMA 2018. LNCS, vol. 10835, pp. 50–61. Springer, Cham (2018). https://doi.org/10.1007/978-3-319-91641-5_5
5. Bujok, P.: Migration model of adaptive differential evolution applied to real-world problems. In: Rutkowski, L., Scherer, R., Korytkowski, M., Pedrycz, W., Tadeusiewicz, R., Zurada, J.M. (eds.) ICAISC 2018. LNCS (LNAI), vol. 10841, pp. 313–322. Springer, Cham (2018). https://doi.org/10.1007/978-3-319-91253-0_30

6. Bujok, P., Tvrdík, J.: Parallel migration model employing various adaptive variants of differential evolution. In: Rutkowski, L., Korytkowski, M., Scherer, R., Tadeusiewicz, R., Zadeh, L.A., Zurada, J.M. (eds.) EC/SIDE -2012. LNCS, vol. 7269, pp. 39–47. Springer, Heidelberg (2012). https://doi.org/10.1007/978-3-642-29353-5_5

7. Bujok, P., Zamuda, A.: Cooperative model of evolutionary algorithms applied to CEC 2019 single objective numerical optimization. In: 2019 IEEE Congress on Evolutionary Computation (CEC), pp. 366–371 (2019). https://doi.org/10.1109/CEC.2019.8790317

8. Das, S., Mullick, S.S., Suganthan, P.N.: Recent advances in differential evolution-an updated survey. Swarm Evol. Comput. **27**, 1–30 (2016)

9. Das, S., Suganthan, P.N.: Problem definitions and evaluation criteria for CEC 2011 competition on testing evolutionary algorithms on real world optimization problems. Technical report, Jadavpur University, India and Nanyang Technological University, Singapore (2010)

10. Das, S., Suganthan, P.N.: Differential evolution: a survey of the state-of-the-art. IEEE Trans. Evol. Comput. **15**, 27–54 (2011)

11. Elsayed, S.M., Sarker, R.A., Essam, D.L.: GA with a new multi-parent crossover for solving IEEE-CEC2011 competition problems. In: 2011 IEEE Congress of Evolutionary Computation (CEC), pp. 1034–1040 (2011)

12. Hansen, N., Kern, S.: Evaluating the CMA evolution strategy on multimodal test functions. In: Yao, X., et al. (eds.) PPSN 2004. LNCS, vol. 3242, pp. 282–291. Springer, Heidelberg (2004). https://doi.org/10.1007/978-3-540-30217-9_29

13. Neri, F., Tirronen, V.: Recent advances in differential evolution: a survey and experimental analysis. Artif. Intell. Rev. **33**, 61–106 (2010)

14. Storn, R., Price, K.V.: Differential evolution - a simple and efficient heuristic for global optimization over continuous spaces. J. Global Optim. **11**, 341–359 (1997)

15. Tang, L., Dong, Y., Liu, J.: Differential evolution with an individual-dependent mechanism. IEEE Trans. Evol. Comput. **19**(4), 560–574 (2015)

16. Tvrdík, J.: Competitive differential evolution. In: Matoušek, R., Ošmera, P. (eds.) MENDEL 2006, 12th International Conference on Soft Computing, pp. 7–12. University of Technology, Brno (2006)

17. Wang, Y., Li, H.X., Huang, T., Li, L.: Differential evolution based on covariance matrix learning and bimodal distribution parameter setting. Appl. Soft Comput. **18**, 232–247 (2014)

18. Wolpert, D.H., Macready, W.G.: No free lunch theorems for optimization. IEEE Trans. Evol. Comput. **1**, 67–82 (1997)

Pareto-Based Self-organizing Migrating Algorithm Solving 100-Digit Challenge

Thanh Cong Truong[1](✉)[iD], Quoc Bao Diep[1][iD], Ivan Zelinka[1][iD],
and Roman Senkerik[2][iD]

[1] Faculty of Electrical Engineering and Computer Science,
VSB-Technical University of Ostrava,
17. listopadu 2172/15, 708 00 Ostrava-Poruba, Ostrava, Czech Republic
{cong.thanh.truong.st,ivan.zelinka}@vsb.cz, diepquocbao@gmail.com
[2] Faculty of Applied Informatics, Tomas Bata University in Zlin,
T. G. Masaryka 5555, 760 01 Zlin, Czech Republic
senkerik@utb.cz

Abstract. In this article, we describe the design and implementation of a variant version of SOMA named SOMA Pareto to solve ten hard problems of the 100-Digit Challenge. The algorithm consists of the following operations: Organization, Migration, and Update. In which, we focus on improving the Organization operation with the adaptive parameters of *PRT* and *Step*. When applying the SOMA Pareto to solve ten hard problems to 10 digits of accuracy, we achieved a competitive result: 85.04 points.

Keywords: Self-organizing migrating algorithm · Optimization function · SOMA Pareto · Swarm intelligence · 100-digit challenge

1 Introduction

The Self-organizing migrating algorithm (SOMA) is a class of swarm heuristic [3], which has been used for stochastic optimization. Ever since first proposed in 1999, SOMA has been used to solve many real domain problems such as engineering application [11], aircraft wing and mechanical part optimization, robot [1], and games [17,18].

Concurrently with technology development, problems arising in practice are increasingly sophisticated and more diverse. Therefore, new improvement for existing algorithms is a need to deal with these new problems. In the case of SOMA, researchers developed different versions of this algorithm to enhance its performance such as CSOMA [12], C-SOMGA [5], M-SOMAQI [13], M-NM-SOMA [14], DSOMA [4], MOSOMA [10], SOMA T3A [8], SOMA Pareto [7] and others. However, to keep pace with the dramatic change of technology, more advanced versions of SOMA are necessary to adapt and solve new problems.

In our previous study [7], we have proposed an algorithm based on SOMA named SOMA Pareto, which proved that it outperformed to previous version of

© Springer Nature Switzerland AG 2020
A. Zamuda et al. (Eds.): SEMCCO 2019/FANCCO 2019, CCIS 1092, pp. 13–20, 2020.
https://doi.org/10.1007/978-3-030-37838-7_2

SOMA. In this work, we uses this algorithm to compete in 100-Digit Challenge competition. This version of SOMA consists of three operations: Organization, Migration, and Update, in which we focus on improving the Organization operation with the adaptive parameters of PRT and $Step$.

The rest of this paper proceeds as follows. Section 2 gives a briefly describes SOMA. Section 3 presents the SOMA Pareto algorithm, which applied to solve the 100-Digit Challenge. Sections 4 and 5 present the experiment setting and results, respectively. Finally, we conclude our work in Sect. 6.

2   The Self-organizing Migrating Algorithm

In this section, we briefly present the SOMA, which is a background for the proposed algorithm later. SOMA [2,15,16] belongs to a class of Swarm Intelligence, a branch of bio-inspired computation based on the emergence of collective intelligence [6] which inspired by the behavior of agents in nature such as ant colonies, honey bees, fireflies, and bird flocks. This algorithm is based on collaborative searching (migrating) the area of all possible solutions (search area). During the search process, the individuals influenced by each other to reach global optima of a problem.

The process starts with randomly generating the individuals in the whole search area. At each iteration, the individual that has the highest fitness value become a Leader and the others become the Migrant (traveling individuals). The Migrant jump step by step in the direction of leader. SOMA algorithm executes in migration loops and in each loop the Migrant travels a certain distance towards the leader in n steps of defined length. Before the traveling individuals jump towards leader, a PRT vector is created. Based on the PRT parameter, the Migrant will jump in the $N - k$ dimensional subspace instead of proceeding directly to the Leader. Equation 1 describes the jumping process.

$$x_{n,j}^{MLnew} = x_{c,j}^{ML} + (x_{L,j}^{ML} - x_{c,j}^{ML})\, t\, PRTVector_j \tag{1}$$

where:

- $t \in\ < 0\ , by$ Step to, PathLength>,
- $x_{n,j}^{MLnew}$: the new position of an individual,
- $x_{c,j}^{ML}$: the current position of an individual,
- $x_{L,j}^{ML}$: the leader position.

3   SOMA Pareto

In this section, we briefly introduce an SOMA-based strategy named SOMA Pareto. Figure 1 illustrates the whole operation of this algorithm. First, a population consist of candidate solutions to the problem is generated. Next, the provided fitness function will evaluates the population. After that, the algorithm executes the migration loop with three primary operations: Organization, Migration, and Update.

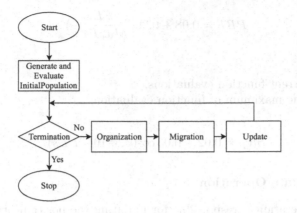

Fig. 1. Flowchart of the algorithm.

3.1   The Organization Operation

The Organization is responsible for choosing the Migrant (traveling individual) and the Leader. The Migrant will then proceed towards the Leader to discover the better locations during the movement.

To select Leader and Migrant, we present a choosing method based on the Pareto Principle, which more detail can be found in [7]. Figure 2 illustrated the process of Organization.

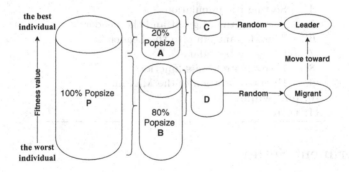

Fig. 2. The organization operation.

3.2   The Migration Operation

The Migration operation is responsible for searching for a better position by moving the Migrant towards the selected Leader. In this algorithm, the authors suggest tuning *PRT* and *Step*, as pointed in Eqs. 2 and 3 to keep maintaining the diversity of the population, and avoiding to be stuck in a local minima.

$$PRT = 0.08 + 0.90(\frac{FEs}{MaxFEs}) \tag{2}$$

where:

- FEs: the current function evaluations,
- $MaxFEs$: the maximum of function evaluations.

$$Step = 0.02 + 0.05\cos(0.5\pi 10^{-6}FEs) \tag{3}$$

3.3   The Update Operation

The Update operation is responsible for updating the position of the Migrant. Technically, a given fitness function will evaluate all positions of the individual while jumping. After that, one of the best position is chosen to compare with the initial position of that Migrant. If the new position is better than the old one, it will substitute the initial position. On the contrary, the Migrant remains unchanged.

Algorithm 1 describe the complete operation of SOMA Pareto [7].

Algorithm 1. SOMA Pareto

1: Generate and evaluate the initial population
2: **while** stopping condition not reached **do**
3:    Update PRT and $Step$
4:    Sorting the population
5:    Selecting the Migrant and the Leader
6:    The Migrant jump to the Leader
7:    Checking boundary
8:    Re-evaluate fitness function
9:    Updated position of the Migrant
10: **end while**
11: **return**

---

4   Experiment Setup

4.1   Hardware and Software Environment

We implemented our SOMA Pareto algorithm in Matlab to optimize ten functions from the 100-Digit Challenge of the SEMCCO 2019 & FANCCO 2019 competition. A full description of all ten test function can be found at: [9]. Table 1 lists the functions used in competition. The experiment is performed on the Intel Core i7 8750H computer with 16 GB RAM, under the Windows 10 64-bit operating system, using Parallel Computing Toolbox of Matlab 2018b version.

Table 1. The 100-Digit challenge basic test functions

No.	Functions	$Fi^* = Fi(x^*)$	Dimensions	Search range
1	Storn's Chebyshev Polynomial Fitting Problem	1	9	$[-8192, 8192]$
2	Inverse Hilbert Matrix Problem	1	16	$[-16384, 16384]$
3	Lennard-Jones Minimum Energy Cluster	1	18	$[-4, 4]$
4	Rastrigin's Function	1	10	$[-100, 100]$
5	Griewangk's Function	1	10	$[-100, 100]$
6	Weierstrass Function	1	10	$[-100, 100]$
7	Modified Schwefel's Function	1	10	$[-100, 100]$
8	Expanded Schaffer's F6 Function	1	10	$[-100, 100]$
9	Happy Cat Function	1	10	$[-100, 100]$
10	Ackley Function	1	10	$[-100, 100]$

4.2 The Parameters

In our experiment, we used three type of parameters as following.

- The fixed parameters that are population size, and the number of jumps of each individuals. The Table 2 present the values of these fixed value.
- The adaptive parameter is *Step* as present in Eq. 3.
- The tuned parameter is PRT as present in Eq. 2, which its value corresponding to each problem are given in Table 3.

Table 2. The fixed parameters

Parameters	Value
The population size	1000
The number of jumps	100

4.3 Stopping Criterion

The evolution will ceases when one of the two following criteria is met:

- Reaching the 10-digit level accuracy.
- Reaching the Maximum number of function evaluations ($MaxFEs = 6 * 10^8$ for all functions).

Table 3. The tuned parameters

Function	PRT
1	$FEs/6 * 10^6$
2	$FEs/6 * 10^8$
3	$FEs/5 * 10^7$
4	$FEs/5 * 10^6$
5	$FEs/10^7$
6	$FEs/3 * 10^6$
7	$FEs/10^8$
8	$FEs/2 * 10^8$
9	$FEs/2 * 10^8$
10	$FEs/10^7$

4.4 How to Score Points

The algorithm will be run 50 consecutive times for every function, and the total number of correct digits in the 25 trials that have the best values will be counted.

For each function, the score is the average number of the correct digits in that 25 best trials. For example, the algorithm reaches more than 50% of the trials achieving all ten digits, then the score of that function is 10 points. The best score for the ten functions in total is 100 points.

5 Simulation Results

Table 4 representative the calculation results for ten functions in the 100-Digit challenge competition. In this table, the first column lists the sequence number of functions. The next columns indicate the number of trials (in total 50 times run) achieved n correct digits, where n = 1 to 10. The final column presents the average number of correct digits in the best 25 runs for each function, which is also the score for that function. The total score (the sum of the scores for all ten functions) is in the bottom-most right-hand of the table.

In this table, our algorithm accomplished 50 over 50 runs achieving ten correct digits from function 1, 2, 3, 4, 5, 6, 7, 10. According to the competition rule as defined in [9], we got the highest score for these functions. In the case of function 8 and 9, we achieved score 2 and 3.04 points, respectively. Accurately, with function 8 we reach two correct digits in 50 times, and there are no times in 50 runs that the algorithm achieves above 3 correct digits. Likewise, with function 9 we had 49 times reaching 3 correct digits, and just 1 out of 50 reaches further correct digits. In total, we achieved the score points 85.04 out of 100.

During the simulation, we discovered that the algorithm stuck in local minima and unable to go beyond the local subspace with function number 8 and function number 9. Furthermore, the function number 2 is more sophisticated than the others, so the algorithm needs more FEs to evaluate so that it can reach ten correct digits.

Table 4. Fifty runs for each function sorted by the number of correct digits

Function	Number of correct digits											Score
	0	1	2	3	4	5	6	7	8	9	10	
1	0	0	0	0	0	0	0	0	0	0	50	10
2	0	0	0	0	0	0	0	0	0	0	50	10
3	0	0	0	0	0	0	0	0	0	0	50	10
4	0	0	0	0	0	0	0	0	0	0	50	10
5	0	0	0	0	0	0	0	0	0	0	50	10
6	0	0	0	0	0	0	0	0	0	0	50	10
7	0	0	0	0	0	0	0	0	0	0	50	10
8	0	0	50	0	0	0	0	0	0	0	0	2
9	0	0	0	49	1	0	0	0	0	0	0	3.04
10	0	0	0	0	0	0	0	0	0	0	50	10
Total:												85.04

6 Conclusion

In this paper, we proposed an algorithm, namely SOMA Pareto, to compete in the 100-Digit Challenge competition. As a result, we got 85.04 points when solving the ten functions of this competition. The SOMA Pareto consists of the Organization, Migration, and Update operations. In which, the crucial factor is the application of the Pareto Principle to choose the Migrant and the Leader for raising the performance of the algorithm. The adaptive *PRT* and *Step* parameters help the algorithm avoid stuck in the local minima and achieving the excellent score in 8 out of 10 functions, except function number 8 and 9. Further research needs to be done to improve the other parameter as well as adapting the *PRT*, and *Step* for a better result in the future.

Acknowledgement. The following grants are acknowledged for the financial support provided for this research: Grant of SGS No. SP2019/137, VSB Technical University of Ostrava. This work was also supported by the Ministry of Education, Youth and Sports of the Czech Republic within the National Sustainability Programme Project no. LO1303 (MSMT-7778/2014), further by the European Regional Development Fund under the Project CEBIA-Tech no. CZ.1.05/2.1.00/03.0089.

References

1. Bao, D.Q., Zelinka, I.: Obstacle avoidance for swarm robot based on self-organizing migrating algorithm. Proc. Comput. Sci. **150**, 425–432 (2019)
2. Davendra, D., Zelinka, I.: Self-organizing migrating algorithm. In: New Optimization Techniques in Engineering. Studies in Computational Intelligence. Springer, Heidelberg (2016). https://doi.org/10.1007/978-3-319-28161-2. https://link.springer.com/book/10.1007%2F978-3-319-28161-2#toc

3. Davendra, D., Zelinka, I., Bialic-Davendra, M., Senkerik, R., Jasek, R.: Discrete self-organising migrating algorithm for flow-shop scheduling with no-wait makespan. Math. Comput. Modell. **57**(1–2), 100–110 (2013)
4. Davendra, D., Zelinka, I., Pluhacek, M., Senkerik, R.: DSOMA—discrete self organising migrating algorithm. In: Davendra, D., Zelinka, I. (eds.) Self-Organizing Migrating Algorithm. SCI, vol. 626, pp. 51–63. Springer, Cham (2016). https://doi.org/10.1007/978-3-319-28161-2_2
5. Deep, K.: Dipti: a self-organizing migrating genetic algorithm for constrained optimization. Appl. Math. Comput. **198**(1), 237–250 (2008)
6. Del Ser, J., et al.: Bio-inspired computation: where we stand and what's next. Swarm Evol. Comput. **48**, 220–250 (2019)
7. Diep, Q., Zelinka, I., Das, S.: Self-organizing migrating algorithm pareto. MENDEL **25**(1), 111–120 (2019)
8. Diep, Q.B.: Self-organizing migrating algorithm team to team adaptive-SOMA T3A. In: 2019 IEEE Congress on Evolutionary Computation (CEC), pp. 1182–1187. IEEE (2019)
9. Price, K.V., Awad, N.H., Ali, M.Z., Suganthan, P.N.: The 100-digit challenge: problem definitions and evaluation criteria for the 100-digit challenge special session and competition on single objective numerical optimization. Technical report. Nanyang Technological University Singapore (2018)
10. Kadlec, P., Raida, Z.: A novel multi-objective self-organizing migrating algorithm. Radioengineering **20**(4), 804–816 (2011)
11. Nolle, L., Zelinka, I., Hopgood, A.A., Goodyear, A.: Comparison of an self-organizing migration algorithm with simulated annealing and differential evolution for automated waveform tuning. Adv. Eng. Softw. **36**(10), 645–653 (2005)
12. dos Santos Coelho, L., Mariani, V.C.: An efficient cultural self-organizing migrating strategy for economic dispatch optimization with valve-point effect. Energy Convers. Manage. **51**(12), 2580–2587 (2010)
13. Singh, D., Agrawal, S.: Hybridization of self organizing migrating algorithm with quadratic approximation and non uniform mutation for function optimization. In: Das, K.N., Deep, K., Pant, M., Bansal, J.C., Nagar, A. (eds.) Proceedings of Fourth International Conference on Soft Computing for Problem Solving. AISC, vol. 335, pp. 373–387. Springer, New Delhi (2015). https://doi.org/10.1007/978-81-322-2217-0_32
14. Singh, D., Agrawal, S.: Nelder-mead and non-uniform based self-organizing migrating algorithm. In: Pant, M., Deep, K., Bansal, J., Nagar, A., Das, K. (eds.) Soft Computing for Problem Solving, vol. 436, pp. 795–807. Springer, Singapore (2016). https://doi.org/10.1007/978-981-10-0448-3_66
15. Zelinka, I.: SOMA-self-organizing migrating algorithm. In: Onwubolu, G.C., Babu, B.V. (eds.) New Optimization Techniques in Engineering, vol. 141, pp. 167–217. Springer, Heidelberg (2004). https://doi.org/10.1007/978-3-540-39930-8_7
16. Zelinka, I., Jouni, L.: SOMA-self-organizing migrating algorithm mendel. In: 6th International Conference on Soft Computing, Brno, Czech Republic (2000)
17. Zelinka, I., Němec, M., Šenkeřík, R.: Gamesourcing: perspectives and implementations. In: Simulation and Gaming. IntechOpen (2017)
18. Zelinka, I., Sikora, L.: StarCraft: brood war–strategy powered by the soma swarm algorithm. In: 2015 IEEE Conference on Computational Intelligence and Games (CIG), pp. 511–516. IEEE (2015)

Population Size in Differential Evolution

Amina Alić[✉], Klemen Berkovič, Borko Bošković[iD], and Janez Brest[iD]

Faculty of Electrical Engineering and Computer Science, Computer Architecture
and Languages Laboratory, Institute of Computer Science, University of Maribor,
2000 Maribor, Slovenia
{amina.alic,klemen.berkovic1,borko.boskovic,janez.brest}@um.si
https://labraj.feri.um.si

Abstract. In this paper we examined how the population size affects
the performance of the differential evolution algorithm. First, we tested
the original differential evolution algorithm, and then the improved self-
adaptive differential evolution algorithm, on ten benchmark functions,
that have been proposed for the CEC 2019 competition. We used six
different population sizes. Afterwards, we tested the newly created algorithm with population reinitialization. The results show that the population size affects the algorithm's efficiency, and that we need to tune it to
obtain the best results. In the paper, we demonstrate that the newly created algorithm with reinitialization gives better, or at least comparable,
results than the two algorithms without reinitialization.

Keywords: Global optimum · Differential evolution ·
Reinitialization · Population size

1 Introduction

Evolutionary computing is a research area inspired by natural evolution [3]. The
main feature of natural evolution is the survival of the fittest. In evolutionary
algorithms, the initial population is generated randomly and the fitness of every
individual is calculated. The best ones survive and reproduce, and so evolution
progresses [3]. Because of simplicity, in evolutionary algorithms, the population
size NP is constant, but we are aware that this is not the case in nature. Instead,
the number of individuals in a population varies in different generations. Adaptive population size is still a challenging task, and, for now, we wanted to see
how the population size affects the algorithms.

We briefly discus the related work in Sect. 2. The original differential evolution algorithm and its improved self-adaptive version are described in Sect. 3.
A new algorithm is proposed in Sect. 4. Section 5 presents our experiments and
results. In Sect. 6 we conclude our work briefly.

The authors acknowledge the financial support from the Slovenian Research Agency
(research core funding No. P2-0041), and the investment co-financed by the Republic of
Slovenia and the European Union, European Regional Development Fund, implemented
under the Operational Program for the Implementation of the EU Cohesion Policy in
the period 2014–2020.

© Springer Nature Switzerland AG 2020
A. Zamuda et al. (Eds.): SEMCCO 2019/FANCCO 2019, CCIS 1092, pp. 21–30, 2020.
https://doi.org/10.1007/978-3-030-37838-7_3

2   Related Work

In the paper [6], the authors investigated how different population sizes affect the differential evolution algorithm. The paper considered the effect of the population sizes 2D, 4D, 6D, 8D and 10D (D - dimension of the problem) with two different mutation strategies on problems chosen from the CEC 2005 Special Session on Real-Parameter Optimization. They found that a smaller population size with a greedy strategy converges fast but premature convergence and stagnation are more pronounced. A large population, with a strategy having good exploration capacity, does not prematurely converge or stagnate but it can converge very slow.

The same authors in another paper [5] have proposed a differential evolution algorithm with an ensemble of parallel populations having different population sizes, in which a more suitable population size takes most of the function evaluations adaptively. Although this paper uses multiple populations, it is related to our work in the manner that it explores how the population size affects the convergence. They found that the multi-population differential evolution algorithm was more efficient in obtaining better quality solutions than the conventional differential evolution algorithm.

A review article on the study of how the population size affects differential evolution [9] emphasizes that the inappropriate choice of the population size may seriously impede the performance of each differential evolution algorithm.

All those papers are considering the adaptation of population size in the conventional differential evolution algorithm, with other parameters (crossover and mutation rates) kept constant. Besides that, we are investigating the effect of changing population size along with adapting other parameters.

3   Background

3.1   Differential Evolution (DE)

In the DE algorithm [2,4,7,8,11,12], population or candidate solutions are represented by real-valued vectors with D components (genes) [3]

$$\boldsymbol{x}_i^{(G)} = x_{i,1}^{(G)}, x_{i,2}^{(G)}, ..., x_{i,D}^{(G)}, \tag{1}$$

where $i = 1, ..., NP$ and NP is the population size. G represents a generation. The offspring are created through the mutation and crossover. The initial population is generated randomly between lower and upper bounds, which are defined by the problem. An evolutionary cycle starts with random selection of 3 vectors $\boldsymbol{x}_{r1}, \boldsymbol{x}_{r2}, \boldsymbol{x}_{r3}$ from the initial population. A mutation vector is then obtained by adding a perturbation vector to the first of those random vectors

$$\boldsymbol{v}_i^{(G+1)} = \boldsymbol{x}_{r1}^{(G)} + \boldsymbol{p}^{(G)}. \tag{2}$$

The perturbation vector p is the difference between two other randomly selected vectors multiplied by the scaling factor F

$$p^{(G)} = F(x_{r2}^{(G)} - x_{r3}^{(G)}). \tag{3}$$

The scaling factor is a positive number, and has values $F \in [0, \infty]$. In most cases, the scaling factor occupies values between $F \in [0, 2]$. The second step in the reproduction is usually a binomial crossover, which has one parameter, the crossover probability $CR \in [0, 1]$. It creates a trial vector, combining elements of the mutation vector and the corresponding parent vector as

$$u_{i,j}^{(G+1)} = \begin{cases} v_{i,j}^{(G+1)}, & \text{if } rand(0, 1) \leq CR \text{ or } j == j_{rand}, \\ x_{i,j}^{(G)}, & \text{otherwise.} \end{cases} \tag{4}$$

CR determines the probability that the trial vector takes a component from the mutation vector, and j_{rand} is a randomly chosen integer in the range $1, ..., D$ which provides that at least one component of the trial vector is changed in regard to the previous generation. A selection of the offspring that will proceed to reproduction in next generation comes at the end of the evolutionary cycle. The fitness value of the trial vector is compared to the fitness value of the previous population member, the parent vector. The fittest member is allowed to be reproduced further

$$x_i^{(G+1)} = \begin{cases} u_i^{(G+1)}, & \text{if } f(u_i^{(G+1)}) \leq f(x_i^{(G)}), \\ x_i^{(G)}, & \text{otherwise.} \end{cases} \tag{5}$$

3.2   Self-adaptive Differential Evolution (jDE)

DE has three parameters, namely F, CR, and NP, and their tuning can improve the performance of the algorithm greatly. In the original DE these parameters are specified before the evolutionary cycle, and remain fixed during each generation of the algorithm. That is in contrast with the dynamic nature of evolutionary computing itself. Furthermore, different values of parameters can be optimal at different stages of the evolutionary process. A better approach is to use self-adapting parameters. In an improved algorithm, that the authors named jDE [1], all population members are extended by the control parameters F and CR. Adaptive changes of the control parameters should give better individuals, in the sense that they will have better fitness values. In jDE new control parameters are calculated as

$$F_i^{(G+1)} = \begin{cases} F_l + rand_1 * F_u, & \text{if } rand_2 < \tau_1 \\ F_i^{(G)}, & \text{otherwise} \end{cases} \tag{6}$$

and

$$CR_i^{(G+1)} = \begin{cases} rand_3, & \text{if } rand_4 < \tau_2 \\ CR_i^{(G)}, & \text{otherwise.} \end{cases} \tag{7}$$

Algorithm 1. Self-Adaptive Differential evolution with Restarts

---
Input: NP, F_l, F_u, τ_1, τ_2, α, ε
Output: Best found solution
1 Initialization;
2 **while** *not stopping criteria met* **do**
3     $count \leftarrow 0$;
4   **for** $i \leftarrow 1$ **to** NP **do**
5   Use Eq. (6) for obtaining new value of F_i;
6   Use Eq. (7) for obtaining new value of CR_i;
7   Use Eq. (2) for creating new mutant vector v_i, where $F = F_i$ and
           $CR = CR_i$;
8   **for** $j \leftarrow 1$ **to** D **do**
9   Use Eq. (4) for crossing over component j;
10   **end**
11   Use Eq. (5) for selection between individual $x_i^{(G)}$ and $x_i^{(G+1)}$;
12   **if** $f(x_b) < f(x_i^{(G+1)})$ **then**
13          $x_b \leftarrow x_i^{(G+1)}$;
14   **end**
15   **if** $|f(x_i^{(G+1)}) - f(x_b)| < \varepsilon$ **then**
16          $count \leftarrow count + 1$;
17   **end**
18   **end**
19   **if** $count \geq NP \cdot \alpha$ **then**
20   Population reinitialization;
21   **end**
22 **end**
23 **return** x_b;

Here, τ_1 and τ_2 represent small probabilities when control parameters should be changed, and $rand_j$, $j = (1, 2, 3, 4)$ are uniform random numbers in the range $[0, 1]$. $F_{i,G+1}$ and $CR_{i,G+1}$ are computed before the mutation, so they have an impact on mutation, crossover and selection operations when making offspring.

4 Self-adaptive Differential Evolution with Restarts (rjDE)

Our main work is based on the newly created algorithm, called rjDE, that is presented in Algorithm 1. The algorithm is designed for tackling CEC 2019 problems proposed in the technical report [10] for the 100-Digit challenge. The proposed rjDE algorithm is derived from the jDE algorithm, so it uses the same technique for control parameters adaptation over each evolutionary step.

The jDE algorithm can have the same problem as the DE algorithm. Both algorithms have fast convergence, and can be trapped into local optima. When the basic DE converges to some local optima, its population diversity is decreased. In order to avoid that, we added Line 15 to Algorithm 1 that checks if the i-th individual fitness value is close to that of the best individual. If this condition is true, we increase the counter. When some individuals obtained similar fitness values as the best one, we can assume that population diversity is also

Table 1. Fifty runs of DE for each function sorted by the number of correct digits.

Function	Number of correct digits											Score
	0	1	2	3	4	5	6	7	8	9	10	
F1	0	0	0	0	0	0	0	0	0	0	50	10
F2	0	0	0	0	0	0	0	0	0	0	50	10
F3	4	42	0	1	0	0	0	0	0	0	3	2.16
F4	48	2	0	0	0	0	0	0	0	0	0	0.08
F5	0	0	35	9	0	0	0	0	0	0	6	4.28
F6	1	34	0	0	0	0	0	0	0	0	15	6.4
F7	49	1	0	0	0	0	0	0	0	0	0	0.04
F8	35	15	0	0	0	0	0	0	0	0	0	0.6
F9	0	18	32	0	0	0	0	0	0	0	0	2
F10	50	0	0	0	0	0	0	0	0	0	0	0
Total:												**35.56**

decreased. Therefore, a reinitialization of the population takes place. A restart in rjDE i.e. population reinitialization, occurs when the fitness values of $\alpha \cdot NP$ individuals differ from the best fitness value by less than a very small value ε.

5   Experiments and Results

We experimented on CEC benchmark functions [10] and followed their rules for computing scores. In [10] there is no limit on the maximum number of function evaluations (*MaxFEs*), but in this work we set $MaxFEs = 10^7$.

We analyzed three different algorithms: DE, jDE, and rjDE on population sizes $NP = 50, 100, 200, 400, 800, 1600$.

We show results for population size 100 first, and, later, compare results for all population sizes, on all 10 functions, including all 3 algorithms.

5.1   Results for $NP = 100$

Table 1 shows the results for all benchmark functions using the original DE algorithm. Other parameters that we used in this algorithm are $F = 0.5$ and $CR = 0.9$. It can be seen that DE obtained 10 correct digits for functions F1 and F2, while for function F10 it obtained no correct digits. Functions F4, F7 and F8 also have scores almost equal to zero, from which we can see that the original DE algorithm is not suitable for solving those functions when using a population with size 100.

The jDE algorithm solved functions F1, F2 and F4 to 10 correct digits. For all other functions, scores were bigger than one correct digit. Those results are shown in Table 2. Parameters used in this algorithm are $F_l = 0.1$, $F_u = 0.9$, $\tau_1 = \tau_2 = 0.1$. Initial control parameters were $F = 0.5$ and $CR = 0.9$. It is obvious that jDE is better for solving single objective optimization problems than the simple DE.

Table 2. Fifty runs of jDE for each function sorted by the number of correct digits.

Function	Number of correct digits											Score
	0	1	2	3	4	5	6	7	8	9	10	
F1	0	0	0	0	0	0	0	0	0	0	50	10
F2	0	0	0	0	0	0	0	0	0	1	49	10
F3	0	39	0	0	0	0	0	0	0	0	11	4.96
F4	2	2	0	0	0	0	0	0	0	0	46	10
F5	0	17	12	0	0	0	0	0	0	0	21	8.88
F6	0	31	0	0	0	0	0	0	0	0	19	7.84
F7	13	31	6	0	0	0	0	0	0	0	0	1.24
F8	0	47	3	0	0	0	0	0	0	0	0	1.12
F9	0	0	49	1	0	0	0	0	0	0	0	2.04
F10	36	0	0	0	0	0	0	0	0	0	14	5.6
Total:												**61.68**

Table 3. Fifty runs of rjDE for each function sorted by the number of correct digits.

Function	Number of correct digits											Score
	0	1	2	3	4	5	6	7	8	9	10	
F1	0	0	0	0	0	0	0	0	0	0	50	10
F2	0	0	0	0	0	0	0	0	0	0	50	10
F3	0	0	2	24	19	0	0	1	0	0	4	5.04
F4	0	0	0	0	0	0	0	0	0	0	50	10
F5	0	0	0	0	0	0	0	0	0	0	50	10
F6	0	0	0	0	0	0	0	0	0	0	50	10
F7	2	3	1	0	0	0	0	0	0	0	44	10
F8	0	48	2	0	0	0	0	0	0	0	0	1.08
F9	0	0	50	0	0	0	0	0	0	0	0	2
F10	0	0	0	0	0	0	0	0	0	0	50	10
Total:												**78.12**

Table 3 shows the score for each function for the rjDE algorithm. The highest score 10 was obtained for 7 out of 10 functions. Parameters used here were the same as for the jDE algorithm, along with two additional parameters, $\varepsilon = 10^{-16}$ and $\alpha = 0.5$. Functions F3, F8, and F9 seem to be the difficult ones for rjDE.

Total scores for DE, jDE and rjDE are 35.56, 61.68 and 78.12, respectively.

5.2 Results for Different Population Sizes

In Table 4 we present the performance of the DE algorithm on all 10 benchmark functions and all 6 population sizes. It is obvious that performance of DE is

Table 4. Scores for each function and different population sizes using DE.

Function	Population size					
	50	100	200	400	800	1600
F1	10	10	10	10	10	10
F2	2.04	10	10	10	10	6
F3	3.52	2.16	1	1.08	1.28	1.52
F4	0.08	0.08	0.52	2.08	0	0
F5	3.16	4.28	3.36	6.04	2.44	1
F6	5.32	6.4	6.76	7.48	10	10
F7	0	0.04	0	0	0.12	0.44
F8	0.2	0.6	1.04	1.2	0.28	0
F9	1.76	2	2	2	2	2
F10	0	0	0.4	2.8	1.56	0
Total	36.08	35.56	35.08	**42.68**	37.68	30.96

not increasing nor decreasing continuously when the population size increases, but we can observe that it has by far the best score for $NP = 400$. Observing all particular functions, it can be seen that the best performance was for the function F1, namely 10 correct digits were obtained for every population size. For the function F2, DE obtained 10 correct digits for population sizes $NP = 100, 200, 400, 800$. Too small and too big population sizes obviously have a bad impact on this function, but too small an NP is still worse than too big. For function F6, the DE algorithm reached 10 correct digits for $NP = 800, 1600$, and all other scores were bigger than 5. Functions F3, F5 and F9 all have scores equal to or greater than 1, while for the remaining functions, zero correct digits were obtained 2 or 3 times. The best total score, equal to 42.68, was obtained for $NP = 400$.

Performance of the jDE algorithm is shown in Table 5. It is obvious that this algorithm has better performance than the previous one. The best score (10 correct digits) was obtained 17 times out of 60 possibilities. For function F1 jDE obtained 10 correct digits for every population size, for F2 and F5 4 times and for F4 3 times. Zero correct digits were obtained only 4 times. The algorithm performed worst on function F7. The best total score, 66.48, was obtained again for population size 400.

Table 6 presents results for the rjDE algorithm. This algorithm reached 10 correct digits 37 times out of 70, which is 52.85% of overall performance. The worst result, zero correct digits, was reached only once, for function F10 and population size 1600. The best total score, 78.6, was obtained for population size of 50.

It is obvious that the population size affects the performance of all differential evolution algorithms. For the jDE and rjDE it seems that the total score increases when we increase the population size until it reaches the maximum

Table 5. Scores for each function and different population sizes using jDE.

Function	Population size					
	50	100	200	400	800	1600
F1	10	10	10	10	10	10
F2	1.92	10	10	10	10	10
F3	2.08	4.96	6.4	7.72	10	0.2
F4	6.76	10	10	10	1.44	0.08
F5	6.92	8.88	10	10	10	10
F6	8.92	7.84	7.12	8.56	9.28	10
F7	0.8	1.24	1.88	0	0	0
F8	1.12	1.12	1	1	0.96	0.2
F9	2.04	2.04	2	2	2	2
F10	5.6	5.6	8	7.2	2.76	0
Total	46.16	61.68	66.40	**66.48**	56.44	42.48

Table 6. Scores for each function and different population sizes using rjDE.

Function	Population size						
	25	50	100	200	400	800	1600
F1	10	10	10	10	10	10	10
F2	0.04	5.08	10	10	10	10	9.36
F3	10	10	5.04	3	2.48	2	1.2
F4	10	10	10	10	10	4.88	2.2
F5	10	10	10	10	10	10	5.96
F6	10	10	10	10	10	10	10
F7	5.88	10	10	8.6	1	0.64	0.04
F8	1.2	1.52	1.08	1	1	1	1
F9	2	2	2	2	2	2	2
F10	10	10	10	10	10	4.16	0
Total	69.12	**78.6**	78.12	74.6	66.48	54.68	41.76

for some NP, and then decreases for bigger population sizes. For the DE algorithm, we can notice a small deviation from that observation. The maximum for different algorithms is obtained for different population sizes: For DE the best population size is 400, followed by 800, for jDE the best population sizes are 400, 200, and for rjDE 50 and 100. Obviously, some algorithms perform better on bigger populations, while the others give better results for smaller population sizes. Mean numbers of restarts for all runs, for each benchmark function and each population size are shown in Table 7. The restarts were most frequent for small

Table 7. Mean number of restarts for 50 runs and different population sizes.

Function	Population size						
	25	50	100	200	400	800	1600
F1	0.16	0.00	0.00	0.00	0.00	0.00	0.00
F2	0.74	0.02	0.00	0.00	0.00	0.00	0.00
F3	0.96	0.16	0.00	0.00	0.00	0.00	0.00
F4	1.06	0.00	0.00	0.00	0.00	0.00	0.00
F5	2.26	0.30	0.00	0.00	0.00	0.00	0.00
F6	0.02	0.00	0.00	0.00	0.00	0.00	0.00
F7	13.72	2.90	0.46	0.02	0.00	0.00	0.00
F8	1.62	0.40	0.00	0.00	0.00	0.00	0.00
F9	0.02	0.04	0.00	0.00	0.00	0.00	0.00
F10	0.04	0.00	0.00	0.00	0.00	0.00	0.00

population sizes $NP = 25, 50$ while for the population sizes $NP = 400, 800, 1600$ there were no restarts at all.

We followed the rules that were suggested for the CEC 2019 competition.

6   Conclusion

We analyzed three algorithms: DE, jDE and rjDE on the CEC 2019 benchmark functions with different population sizes, in order to see how the population size affects their performance. We followed the rules of the CEC 2019 competition. Our analysis shows that the population size affects the performance of those algorithms in the manner that it increases the total score until it reaches maximum. Further increment of the population size decreases the total score. The self-adaptive differential evolution with reinitialization has proven to have the best results when performing on selected benchmark functions. For the future work, we plan to run the algorithms with a greater maximum number of function evaluations. We also plan to investigate linear population reduction methods such as L-SHADE [13].

References

1. Brest, J., Greiner, S., Bošković, B., Mernik, M., Žumer, V.: Self-adapting control parameters in differential evolution: a comparative study on numerical benchmark problems. IEEE Trans. Evol. Comput. **10**(6), 646–657 (2006)
2. Das, S., Mullick, S.S., Suganthan, P.N.: Recent advances in differential evolution-an updated survey. Swarm Evol. Comput. **27**, 1–30 (2016)
3. Eiben, A.E., Smith, J.E.: Introduction to Evolutionary Computing. Natural Computing. Springer, Heidelberg (2003). https://doi.org/10.1007/978-3-662-05094-1

4. Eltaeib, T., Mahmood, A.: Differential evolution: a survey and analysis. Appl. Sci. **8**(10), 1945 (2018)
5. Mallipeddi, R., Suganthan, P.: Differential evolution algorithm with ensemble of populations for global numerical optimization. Opsearch **46**(2), 184–213 (2009)
6. Mallipeddi, R., Suganthan, P.N.: Empirical study on the effect of population size on differential evolution algorithm. In: 2008 IEEE Congress on Evolutionary Computation (IEEE World Congress on Computational Intelligence), pp. 3663–3670. IEEE (2008)
7. Maučec, M.S., Brest, J.: A review of the recent use of differential evolution for large-scale global optimization: an analysis of selected algorithms on the CEC 2013 LSGO benchmark suite. Swarm Evol. Comput. (2018, On line). https://doi.org/10.1016/j.swevo.2018.08.005
8. Neri, F., Tirronen, V.: Recent advances in differential evolution: a survey and experimental analysis. Artif. Intell. Rev. **33**(1–2), 61–106 (2010)
9. Piotrowski, A.P.: Review of differential evolution population size. Swarm Evol. Comput. **32**, 1–24 (2017). https://doi.org/10.1016/j.swevo.2016.05.003
10. Price, K.V., Awad, N.H., Ali, M.Z., Suganthan, P.N.: Problem definitions and evaluation criteria for the 100-digit challenge special session and competition on single objective numerical optimization. Technical report, Nanyang Technological University, Singapore, November 2018. http://www.ntu.edu.sg/home/epnsugan/
11. Qin, A.K., Huang, V.L., Suganthan, P.N.: Differential evolution algorithm with strategy adaptation for global numerical optimization. IEEE Trans. Evol. Comput. **13**(2), 398–417 (2009)
12. Storn, R., Price, K.: Differential evolution - a simple and efficient heuristic for global optimization over continuous spaces. J. Global Optim. **11**, 341–359 (1997)
13. Tanabe, R., Fukunaga, A.S.: Improving the search performance of shade using linear population size reduction. In: 2014 IEEE Congress on Evolutionary Computation (CEC), pp. 1658–1665 (2014)

Channel Assignment with Ant Colony Optimization

Marko Peras[✉] and Nikola Ivkovic

Faculty of Organization and Informatics, University of Zagreb,
Pavlinska 2, 42000 Varazdin, Croatia
{marko.peras,nikola.ivkovic}@foi.hr

Abstract. In wireless communication arise various forms of optimization problems including channel assignment problem. There are many possible ways to assign channels to wireless links and our goal is to find the assignment that minimizes channel interference. For that purpose we have developed an ant colony optimization algorithm based on general guidelines of MAX-MIN Ant System and implemented it in C++ language. The algorithm was tested on problem instances and the results showed that the proposed algorithm is learning about instances that is solving and that way improves solution quality with the increase of iterations. Results confirmed that the proposed algorithm is an appropriate approach for solving channel assignment problem in cellular networks.

Keywords: Wireless networks · Combinatorial optimization ·
Computational intelligence · Metaheuristics

1 Introduction

Combinatorial optimization problems arise in many practical applications. Although modern computers can perform computations very fast, some problems like those belonging to NP-hard class still cannot be solved by exact algorithms, unless for very small problem instance or some special cases. Therefore heuristical algorithms are commonly used since they are often able to find good solutions in a reasonable time. In wireless networking, it is important to provide a good quality of service that includes communication with minimal interference. It is also beneficial to maximize the number of available channels, minimize the required number of frequencies, etc. There are various versions of channel assignment problem (CAP) and they can be challenging to solve.

Computational intelligence algorithms are successfully applied for versatile demanding optimization problems. They use heuristical approaches and learning techniques to solve problems. Ant colony optimization is a swarm intelligence approach where artificial ants construct solutions. Their construction is guided by pheromone trails associated with solution components. Pheromone trails are maintained to exploit experience gained by previous successful solutions. Information about less successful solutions is filtered out by forgetting mechanism: pheromone evaporation. ACO efficiently solves problems similar to CAP (QAP).

© Springer Nature Switzerland AG 2020
A. Zamuda et al. (Eds.): SEMCCO 2019/FANCCO 2019, CCIS 1092, pp. 31–42, 2020.
https://doi.org/10.1007/978-3-030-37838-7_4

In the paper, we present an ant colony optimization algorithm for a variant of channel assignment problem where the goal is to assign channels to links in a way that overall interference is minimized. Our algorithm learns about the problem that is being solved and provides promising results.

Heuristic and approximate approaches are rather popular for solving various channel assignment problems, also known as frequency assignment problem. Luna et al. proposed $(1, \lambda)$ evolutionary algorithm for real-world GSM network [10]. Optimization of deployment parameters for indoor office wireless local area network based on IEEE 802.11 standard was presented in [3]. A multi-objective genetic algorithm hybridized with game theory ideas for CAP was studied in [8]. Shukla et al. proposed a heuristic algorithm that uses a technique of increasing the co-site interference depending upon the adjacent channels interferences [15]. Ghosal and Ghosh proposed a differential coloring technique using prediction based and random coloring [4]. Sharma and Chaudhari proposed a way to reduce CAP to satisfiability using graph k-colorability and solve it using propositional satisfiability (3-CNF-SAT) [14]. Leu and Liu proposed concentric hexagon oriented multi channel assignment (CHOMA) with APs and channels arranged in concentric hexagon groups to reduce interference among APs [9]. Buttar, Goel and Kumar propose an algorithm based on wild dogs using intelligent strategies when hunting their prey [1]. Marappan and Sethumadhavan proposed solving channel allocation problem using new genetic algorithm with clique partitioning method [11]. Peter and Olusegun proposed a scheme combining neural network easy convergence ability and genetic algorithm global search ability for optimal dynamic channel assignment in mobile networks [13]. Valdivieso et al. proposed CAP solution based on a centralized simulated annealing algorithm [16]. Kari, Shashidhar and Kentros propose online soft edge coloring model for solving CAP in wireless networks [7]. Novillo, Valdivieso and Velasquez propose a centralized simulated annealing algorithm (CSA) in opportunistic spectrum access WLAN with channel prioritization and channel bandwidth restrictions [12]. Chatterjee and Das proposed ACO variant for routing in mobile Ad-hoc network [2].

The rest of this paper is structured as follows. Section 2 explains the variant of channel assignment problem (CAP) that is studied in this research, followed by description of our algorithm for CAP based on ant colony optimization metaheuristic in Sect. 3. Experimental results are presented in Sect. 4 and final conclusions are given in Sect. 5.

2   Channel Assignment Problem

In various wireless communication applications, in military, satellite, wireless local area networks, TV and radio broadcasting, and cellular mobile networks there are hard optimization problems related to channel assignment or frequency assignment. This study focuses on channel assignment in cellular mobile networks.

In cellular mobile networks, each mobile network provider has a limited frequency band $[f_{min}, f_{max}]$ assigned by the authority in a particular country. This frequency band is divided into N channels of the same width (difference between frequencies of adjacent channels). Hence, the set of channels assigned to provider can be marked $C = \{1, ..., N\}$. To establish a two-way communication between two points, two channels sufficiently separated must be used per connection. Usually, frequency band is divided into two sub-bands $C_1 = \{1, ..., N/2\}$ and $C_2 = \{1 + s, ..., N/2 + s\}$ where s is a distance needed to avoid interference.

Provider positions base stations (cell towers) in a way that enables efficient communication between adjacent base stations and to maximize base station area of coverage. Each base station (i.e. cell) uses a set of channels (frequencies) that can be used to establish a connection. Number of channels assigned to a cell is estimated by the demand of that area. More densely populated areas require a higher number of channels per cell as more people may communicate at the same time.

As available frequency band is limited, with increase of number of communicating devices, probability of interference occurring between two connections increases. Interference between two connections occurs if two connections use the same or near frequency for communication and two connections are geographically close to each other. Interference will occur at the place where transmitting energy is approximately the same for both signals.

Mobile network operators have to distribute available channels among cells in their network in such a way that: 1. each cell receives enough channels to satisfy the demand for that area 2. each channel in a cell is sufficiently separated from other channels within the same cell to avoid interference 3. each channel in a cell is sufficiently separated from channels used in adjacent cells to avoid interference.

A measure of separation between two channels is called distance between two channels. E.g. if two channels have a distance of 3, this means they are separated by 3 channel widths ($3 \times \Delta$). If first channel operates at 800 MHz and Δ is 20 MHz, the second would operate at 860 MHz. Distance between channels within the same cell must be greater than distance between channels in adjacent cells. As power of the cell signal decreases with distance, distance between channels used in two remote cells can be smaller. If two cells are sufficiently distanced from each other, the same channel can be used in these cells. In practical applications, the required distance between channels can be calculated from measurement of the power of cell signal at various locations.

It is not always possible to find an arrangement of channels among cells to completely avoid interference. This depends on number of cells within a network, distances between cells and number of channels available. The studies have shown, that channel assignment problem belongs to group of NP complete problems. Mobile network operators have devised a number of strategies to achieve better results. The goals can be to minimize interference, maximize the number of available links, minimize the number of required channels, minimize required frequency band etc.

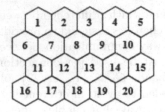 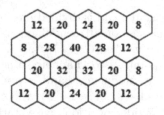

Fig. 1. Cell tower arrangement and demand (nr. of links) per cell

Our study is focused on a variant of CAP where for a given set of cells for mobile communication and the set of available channels the goal is to find an arrangement for which total interference in the system is minimized.

Problem can be formulated as a set of channels $C = \{1, ..., N\}$ that can be assigned to links in mobile network cells. Each cell within a network contains a number of links that corresponds to demand for that area. E.g. if demand for a cell 1 is 12 this cell will contain 12 links, if demand for a cell 2 is 20 this cell will contain 20 links. In this way, demand for each cell is introduced in model. Figure 1 displays an example of cell tower arrangement and demand per cells (instance s400). To avoid interference, channels have to be assigned to links in such a way that channel distance constraints are respected: for channels in the same cell, distance must be greatest, and for neighboring cells distance is lowered as cells are farther apart. In our example, channels assigned to the same cell should be separated by at least 5 channels; channels assigned to adjacent cells should be separated by at least 2 channels, and channels assigned to a second and third ring of cells should be separated by at least 1 channel. (i.e. they must not be the same). Other cells may use the same channel, i.e. distance is 0. Figure 2 shows distances for cell 7. Channels assigned to cell 7 have to be separated by a distance of 5, channels in adjacent cells (1, 2, 8, 12, 11, 6) have to be separated by a distance of 2, channels in second ring of cells (3, 9, 13, 18, 17, 16) have to be separated by a distance of 1, channels in third ring of cells (4, 10, 14, 19) have to be separated by a distance of 1 and channels in all other cells (5, 15, 20) may use the same channels as cell 7 (distance between channels is 0).

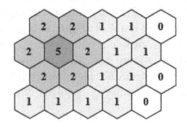

Fig. 2. Required distances between channels assigned to cell links for cell 7

For a two-way connection, pairs of channels with a fixed distance s are used which ensures that there is no interference between two channels used in the same connection. Hence, it is enough to determine one channel $x \in C_1$ per connection as the other channel is calculated as $x + s$.

In our study, the problem is given by following input data: number of cells, with given demand vector, number of available channels and channel distance vector D. Input data produces channel distance matrix derived from cell locations and respective channel distances.

Our study uses modified quadratic assignment problem (QAP) approach using MAX-MIN ants system (MMAS) to find an arrangement of channels per cells such that total interference in the system is minimized. Our study uses approach similar to QAP by allocating channels to links, but differs in a way that there may be more or less channels than links available and the same channel may be reused by two or more links.

To demonstrate how algorithm works, let's assume that we have a small problem with 6 cells, with demand vector V = 1, 2, 1, 1, 2, 1 which produces 8 links, 6 available channels F = 1, 2, 3, 4, 5, 6 and distance vector D = 2, 1, 0. Let cells be arranged in following way (Fig. 3):

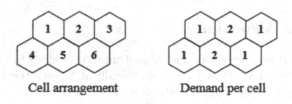

Cell arrangement   Demand per cell

Fig. 3. Cell arrangement and demand per cell of a small CAP

This results in link distance matrix 8×8 presented in Table 1. From this matrix we can read that channel in link 1 (row 1) has to be at least 1 unit

Table 1. Small CAP channel distance matrix

–	Cells	1	2	2	3	4	5	5	6
Cells	Links	1	2	3	4	5	6	7	8
1	1	2	1	1	0	1	1	1	0
2	2	1	2	2	1	0	1	1	1
2	3	1	2	2	1	0	1	1	1
3	4	0	1	1	2	0	0	0	1
4	5	1	0	0	0	2	1	1	0
5	6	1	1	1	0	1	2	2	1
5	7	1	1	1	0	1	2	2	1
6	8	0	1	1	1	0	1	1	2

distanced from channels in links 2 and 3 (columns 2 and 3); and 0 units distanced from channel in link 4 (column 4), i.e. the same channel may be used. We can also see that channel in link 2 (row 2) has to be at least 2 units distanced from channel in link 3 (column 3) which resides in the same cell (cell 2).

For a problem of such small scale, MMAS quickly finds an optimal solution where total interference in the system equals 0 (Fig. 4).

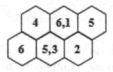

Fig. 4. Optimal solution (with no interference) of a simple problem

Complex problems contain large number of links and channels. With number of links and channels, number of possible solutions increases exponentially. In some instances, a solution with no interference cannot be found. In these cases, we aim to find a solution with minimum interference.

3   Proposed Algorithm

To deal with CAP we have devised an algorithm that belongs to the class of ant colony optimization (ACO) and more specifically if follows the general framework of MAX-MIN ant system (MMAS) [5,6]. The algorithm iteratively performs procedures as described in Algorithm 1. By using pheromone trails and heuristic information, ants construct solutions in each iteration of the algorithm. This is followed by pheromone evaporation and pheromone reinforcement. The algorithm goes through multiple iterations until stopping criteria are not satisfied.

Algorithm 1. The General Procedures of Proposed Algorithm

1: INITIALIZATION()
2: **while** (number of steps is not reached) and (threshold value is not reached) **do**
3:     **for** $k = 1$ to m **do**
4:         CONSTRUCT SOLUTION()
5:     **end for**
6:     PERFORM LOCAL OPTIMIZATION()   ▷ optional
7:     EVAPORATE PHEROMONE TRAILS()
8:     REINFORCE PHEROMONE TRAILS()
9: **end while**

Pheromone trail is associated with pair (link, channel). The pheromone trails can be conveniently stored into a matrix of dimension (number of links) × (number of channels). Each channel can be assigned to any link and even to multiple links. The heuristic information is calculated dynamically whenever channel j is considered to be assigned to link i by using expression (1). The interference between pair (k, l) and (i, j) is denoted by function $interference(c_{kl}, c_{ij})$ and S^P is partial solution that contains all already selected solution components. The pheromone trail τ_{ij} and heuristic information η_{ij} are used by random-proportional rule of MMAS defined by expression (2) to add solution component into partial. The set C contains all available channels.

$$\eta_{ij} = \frac{1}{1 + \sum_{c_{kl} \in S^P} interference(c_{kl}, c_{ij})} \tag{1}$$

$$p_{ij} = \frac{\tau_{ij}^{\alpha} \cdot \eta_{ij}^{\beta}}{\sum_{i,k} \left(\tau_{ik}^{\alpha} \cdot \eta_{ik}^{\beta} \right)}, \forall k \in C, \tag{2}$$

After ants have finished with solution construction, the pheromone trails of all components are evaporated by multiplying with $(1 - \rho)$ as defined with expression (3). The parameter ρ is evaporation rate chosen to be from interval $[0, 1\rangle$, L is the set of links and C is the set of channels. The value of τ_{ij} must not fall below the lower pheromone bound τ_{min}.

$$\tau_{ij} = \max\{\tau_{min}, (1 - \rho) \cdot \tau_{i,j}\}, \forall i \in L, \forall j \in C, \tag{3}$$

The pheromone evaporation is followed by the pheromone reinforcement procedure where only the pheromone trails τ_{kl} of the best solution are increased by adding additional value to a pheromone trails by using expression (4)

$$\tau_{kl} = \min\{\tau_{max}, \tau_{kl} + \frac{1}{\rho}\}, \forall \tau_{kl} \in s^{best} \tag{4}$$

The initial value for all pheromone trails τ_0 and the upper pheromone bound τ_{max} are set to equal value by using expression (5).

$$\tau_0 = \tau_{max} = \frac{1}{\rho} \tag{5}$$

4   Experimental Research

4.1   Experimental Settings

In order to evaluate and analyze behaviour of our algorithm we have performed experimental research. For testing purposes we have used eight problem instances listed in Table 2. These problems model mobile networks with 20 cells but with different number of links per cell, different number of total links and different number of available channels that can be assigned to links.

For this research we have implemented proposed algorithm in C++11 language and compiled the program with g++ compiler under Linux environment. For each problem instance we have repeated experiment eleven times and calculated statistical values. For all reported results in the following subsection, except in few cases when it is explicitly specified otherwise, we have used the same parameter setting. Each algorithm was allowed to execute 2000 iterations, each colony had 50 ants, other parameters was set to $\alpha = 1$, $\beta = 4$, $\rho = 0.2$, $\tau_0 = 1/\rho$, $\tau_{max} = 1/\rho$, $\tau_{min} = 0.001$.

Table 2. Problem instances used in this research

Problem instance	Number of cells	Links in the smallest cell	Links in the largest cell	Total num. of links	Num. of available channels
t100f65	20	5	5	100	65
s100f80	20	3	10	100	80
t200f130	20	10	10	200	130
s200f160	20	4	20	200	160
t300f180	20	15	15	300	180
s300f240	20	6	28	300	240
t400f240	20	20	20	400	240
s400f310	20	8	40	400	310

4.2 Results

At the beginning, the algorithm is not expected to find good solutions unless the problems are trivial, but it is interesting to observe some statistics in order to compare it with the results at the end of execution. Some basic statistics about observed best solutions in 10th iteration of the algorithm is presented in Table 3. For each problem instance minimum solution, maximum solution, arithmetic mean, and median solution are provided. This data shows that for smaller problem instances, with smaller number of cells and links, cumulative interference is smaller and it generally grows with the size of the problem.

To investigate the benefits of learning mechanism in the proposed algorithm, two typical executions on t100f65 are presented in Fig. 5. In the case where parameter α is set to 0, which effectively disables the learning mechanism, the algorithm randomly fluctuates with solution quality mostly between 250 and 300. For the case when parameter α is set to 1, the algorithm achieves good progress from solution with fitness around 300 towards a solution of fitness 10. As expected, at the beginning, the best solution in current iteration fluctuates similar to the case without learning, but as the pheromone values accumulate more information about successful solutions in the previous iterations these fluctuations are much smaller.

Although, the proposed algorithm can achieve rather good results for smaller problem instances even without additional guidance provided by heuristic information η, as in case in Fig. 5, where heuristic information was disabled by setting $\beta = 0$, for larger instances the algorithm converges too slowly. This is shown in Fig. 6. When parameter $\beta = 0$, algorithm does improve solution quality with increase of algorithm iterations, but this becomes too slow and does not reach good quality solutions. When provided with $\alpha = 1$ and $\beta = 4$, the algorithm starts with better solutions and converges much faster and closer to optimal (in this case also the ideal) solution.

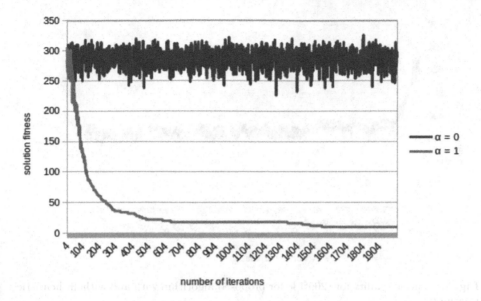

Fig. 5. Solution quality for t100f65 with parameter $\beta = 0$

The final results after 2000 iterations are presented in Table 4. In addition to values provided in Table 3, 0.2-quantile and the percentage of executions that obtained ideal solutions are also reported. The results show that for most problem instances the proposed algorithm did find ideal solutions in some executions of the algorithm, and depending on the problem instance, this happened between 18% to 46% of repetitions. The average performance of the algorithm in those cases is such that arithmetic mean and median solution are rather close to ideal solution. Although in one execution of the algorithm it is not expected to obtain ideal solution, by repeating the algorithm multiple times it is reasonable to expect to obtain an ideal solution. For problem instances t300f180, t400f240, and s400f310 the algorithm did not obtain any ideal solution. Although it is important to note that optimal solutions are not always equal to ideal solutions because due to available channels and required constraints ideal solutions are sometimes impossible. For those harder problem instances, improvement of arithmetic mean of best obtained solution with regard to iterations is presented in Fig. 7. It is evident that the algorithm slows down with improvement after around 100 iterations, but the improvement in solution quality continues until the end of execution suggesting that for more iterations algorithm might improve solution further. In comparison to results from the beginning of the algorithm, the final results show considerable improvement.

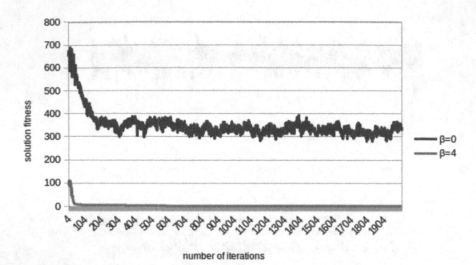

number of iterations

Fig. 6. Solution quality for t200f130 for proposed algorithm with and without heuristic information

Table 3. Solutions obtained at the beginning of the algorithm (after 10 iterations)

Problem instance	MIN	MAX	MEAN	MEDIAN
t100f65	20	38	32.2	33
s100f80	30	44	38.2	39
t200f130	72	92	85.8	87
s200f160	82	106	100.94	98
t300f180	164	204	186.8	188
s300f240	124	156	146.4	149
t400f240	248	266	260.2	264
s400f310	224	258	242.4	244

Table 4. The final results obtained after 2000 iterations

Problem instance	MIN	MAX	MEAN	MEDIAN	$Q_{0.2}$	Ideal solutions
t100f65	0	4	1.4	1	0	45.5%
s100f80	0	4	1.6	2	0	27.3%
t200f130	0	4	1.8	2	1.6	18.2%
s200f160	0	6	3.4	4	1.6	18.2%
t300f180	18	32	27	27	24	0.0%
s300f240	0	6	2.6	2	1.6	18.2%
t400f240	20	40	27.8	27	21.6	0.0%
s400f310	10	16	14.6	16	13.6	0.0%

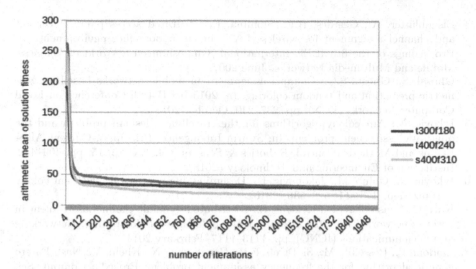

Fig. 7. Arithmetic mean of best found solutions for harder problem instances

5 Conclusion

Optimization in wireless communication is important to keep high quality of service, minimize resource usage and allow as many of serviced users as possible. To solve a variant of channel assignment problem (CAP) we have devised and implemented an ant colony optimization algorithm. Although the problem seems similar to the quadratic assignment problem (QAP), because of different objectives and different constraints our design choices are somewhat different from those usually used in ACO for QAP. In ACO for QAP the heuristic information are not commonly used, but in our algorithm for CAP heuristic information showed to be an important component in order to cope with problem instances of greater size. In performed experimental research our algorithm demonstrated learning capabilities and proved to be a viable approach in solving CAP. In future work our plans are to add local optimization and perform parameters exploration to improve the algorithmic behaviour as well as comparison with other channel assignment schemes based on GA, PSO and other metaheuristics.

References

1. Buttar, A.S., Goel, A.K., Kumar, S.: Solving 55-cell benchmark frequency assignment problem by novel nature inspired algorithm. In: 2014 International Conference on Signal Processing and Integrated Networks (SPIN), pp. 407–411, February 2014
2. Chatterjee, S., Das, S.: Ant colony optimization based enhanced dynamic source routing algorithm for mobile ad-hoc network. Inf. Sci. **295**, 67–90 (2015)

3. Eisenblätter, A., Geerdes, H.F., Siomina, I.: Integrated access point placement and channel assignment for wireless LANs in an indoor office environment. In: Proceedings of the 8th IEEE International Symposium on a World of Wireless, Mobile and Multimedia Networks, June 2007
4. Ghosal, S., Ghosh, S.C.: Channel assignment in mobile networks based on geometric prediction and random coloring. In: 2015 IEEE 40th Conference on Local Computer Networks (LCN), pp. 237–240, October 2015
5. Ivkovic, N.: Ant colony algorithms for the travelling salesman problem and the quadratic assignment problem. In: Swarm Intelligence: Principles, Current Algorithms and Methods, Control, Robotics & Sensors, vol. 1, chap. 15, pp. 409–442. Institution of Engineering and Technology (2018)
6. Ivkovic, N., Golub, M., Jakobovic, D.: Designing dna microarrays with ant colony optimization. JCP **11**, 528–536 (2016)
7. Kari, C., Shashidhar, N., Kentros, S.: Distributed dynamic channel assignment in wireless networks. In: 2014 International Conference on Computing, Networking and Communications (ICNC), pp. 1113–1117, February 2014
8. Laidoui, F., Bessedik, M., Si-Tayeb, F.B., Bengherbia, N., Khelil, Y.: Nash-Pareto genetic algorithm for the frequency assignment problem. Procedia Comput. Sci. **126**, 282–291 (2018). Knowledge-Based and Intelligent Information & Engineering Systems: Proceedings of the 22nd International Conference, KES-2018, Belgrade, Serbia
9. Leu, F., Liu, P.: A channel assignment and AP deployment scheme for concentric-hexagon based multi-channel wireless networks. In: 2010 13th International Conference on Network-Based Information Systems, pp. 504–509, September 2010
10. Luna, F., Alba, E., Nebro, A.J., Pedraza, S.: Evolutionary algorithms for real-world instances of the automatic frequency planning problem in GSM networks. In: Cotta, C., van Hemert, J. (eds.) Evolutionary Computation in Combinatorial Optimization, pp. 108–120. Springer, Heidelberg (2007). https://doi.org/10.1007/978-3-540-71615-0_10
11. Marappan, R., Sethumadhavan, G.: Solving channel allocation problem using new genetic algorithm with clique partitioning method. In: 2016 IEEE International Conference on Computational Intelligence and Computing Research (ICCIC), pp. 1–4, December 2016
12. Novillo, F., Valdivieso, C., Velasquez, F.: Centralized channel assignment algorithm for OSA-enabled WLANs based on simulated annealing. In: 2015 7th IEEE Latin-American Conference on Communications (LATINCOM), pp. 1–6, November 2015
13. Peter, E.U., Olusegun, A.O.: A neural network and genetic algorithm scheme for optimal dynamic channel assignment in mobile networks. In: 2017 IEEE 3rd International Conference on Electro-Technology for National Development (NIGER-CON), pp. 139–144, November 2017
14. Sharma, P.C., Chaudhari, N.S.: Channel assignment problem in cellular network and its reduction to satisfiability using graph k-colorability. In: 2012 7th IEEE Conference on Industrial Electronics and Applications (ICIEA). pp. 1734–1737, July 2012
15. Shukla, A., Tiwari, R., Rungta, S., Kumar, M.S.: A new heuristic channel assignment in cellular networks. In: 2009 WRI World Congress on Computer Science and Information Engineering, vol. 7, pp. 473–478, March 2009
16. Valdivieso, C., Novillo, F., Gomez, J., Dik, D.: Centralized channel assignment algorithm for WSN based on simulated annealing in dense urban scenarios. In: 2016 8th IEEE Latin-American Conference on Communications (LATINCOM), pp. 1–6, November 2016

Self-organizing Migrating Algorithm with Non-binary Perturbation

Michal Pluhacek[⊠], Roman Senkerik, Adam Viktorin,
and Tomas Kadavy

Faculty of Applied Informatics, Tomas Bata University in Zlin,
Nam T.G. Masaryka 5555, 760 01 Zlin, Czech Republic
{pluhacek, senkerik, aviktorin, kadavy}@utb.cz

Abstract. The self-organizing migrating algorithm (SOMA) is a popular population base metaheuristic. One of its key mechanisms is a perturbation of the individual movement with a binary-valued perturbation (PRT) vector. The goal of perturbation is to improve the diversity of the population and exploration of the search space. In this paper, we study a variant of the SOMA algorithm with non-binary PRT vector. We investigate the effect of introducing a third possible value, a negative (repulsive) element, into the PRT vector. The aim is to slow the population convergence and prolong the exploration phase. The inspiration is taken from previous successful implementations of repulsive mechanics in another swarm-based method: the Particle Swarm Optimization.

Keywords: Self-organizing migrating algorithm · SOMA · Repulsivity · Perturbation

1 Introduction

Swarm intelligence based metaheuristics [1] such as Particle swarm optimization [2] and Ant colony optimization [3] have gained significant popularity in the past decades and have inspired numerous new bio-inspired algorithms, especially in the field of continuous optimization [4]. The usefulness of these methods is that a solution for very complex optimization problems can be obtained in a reasonable time. As new and more complex optimization challenges emerge daily, the demand for new and more powerful optimizers is constant.

The self-organizing migrating algorithm (SOMA) is a population-based metaheuristic that was originally proposed in 2000 [5]. Subsequently, several strategies for population migration were proposed [6]. Over time, the "All-to-One" and "All-to-All" strategies became the most popular and widely used [7]. SOMA has been successfully applied in various areas [8–10]. Other researchers proposed hybrid variants of SOMA, e.g. the C-SOMGA [11] a combination of SOMA and genetic algorithm (GA) to solve constrained nonlinear optimization problems. Another hybrid of SOMA and GA was proposed in [12].

More recently, SOMA has been used in obstacle avoidance for swarm robot [13], noise removal [14] or pupil localization [15]. Further, SOMA has been subject to recent

A. Zamuda et al. (Eds.): SEMCCO 2019/FANCCO 2019, CCIS 1092, pp. 43–57, 2020.
https://doi.org/10.1007/978-3-030-37838-7_5

theoretical studies such as [16] where the centrality measures were employed to compare various runs of SOMA algorithm.

One of the main inner mechanisms of SOMA is so-called perturbation. It fulfills the role of random mutation, similarly to other methods. A random vector of zeros and ones multiplies the movement-guiding vector of an individual, effectively blocking movement in random dimensions. The PRT parameter controls the probability of zero and one. This mechanism is mainly aimed to improve the diversity of the population and prevent premature convergence into local sub-optima, and it does succeed in fulfilling this role, to some degree [17].

One of the promising methods to further improve population diversity in swarm methods is a repulsive mechanism [17, 18]. In the repulsive strategy, the general guidance vector of an individual is multiplied by a negative number or vector of numbers, leading to movement from the given attraction point (varies by the algorithm). In [7], the negative value for PRT vector is discussed as a possibility. However, no study on this matter has been published.

Thus, the original contribution of this paper can be summarized as:

- Investigation of the possibility of the repulsive mechanism implementation into SOMA algorithm and presentation of related experiments results.
- Detailed numerical as well as graphical analyses of the above-mentioned case study.
- A total of 209 combinations of control parameters settings have been tested.

The rest of the paper is structured as follows: In the following section, the SOMA algorithm is introduced. The experiment methodology is described in section three, and the results presented in the following section. The paper concludes with the discussion of the results.

2   Self-organizing Migrating Algorithm (SOMA)

SOMA is a population-based metaheuristic method that utilizes the traditional crossover and mutation operators in a modified manner simulating a social group of individuals.

In the original and most common SOMA variant (strategy) called All-to-One [6, 7], utilized in this study, the algorithm follows these steps: At the start of each iteration (called migration loop; ML), a Leader is elected based on the cost function (CF) value (individual with the lowest CF value becomes the leader). Each remaining individual then moves in the direction towards the Leader in the search space. The movement consists of jumps determined by the Step parameter until the individual reaches the final position given by the *PathLength* parameter.

Each step is evaluated using the cost function, and the best position (including the initial position of a given individual) is chosen as the new position of the individual in the next migration loop. The exact position of each step is calculated according to (1)

$$x_{i,j}^{ML+1} = x_{i,j,START}^{ML} + (x_{L,j}^{ML} - x_{i,j,START}^{ML}) \cdot t \cdot PRTVector_j \tag{1}$$

Where:

$x_{i,j}^{ML+1}$-value of i-th individual's j-th parameter, in step t in migration loop $ML + 1$,
$x_{i,j,START}^{ML}$-value of i-th individual's j-th parameter, Start position in actual ML,
$x_{L,j}^{ML}$-value of Leader's j-th parameter in migration loop ML,
t - $step \in$ <0, by $step$ to, $PathLength$>,

PRTVector (*Perturbation Vector*) represents the D-dimensional vector of ones and zeros dependent on user predefined *PRT* value. This value can be understood as a threshold constant. For each dimension component in the perturbation vector, a random number r from interval <0, one> is generated. If $r < PRT$, then one is saved to *PRTVector*. Otherwise, zero is saved into the *PRTVector*.

New *PRTVector* is constructed before each step of a given individual, and further, as a rule, there has to be at least one zero in the vector. An example of possible individual trajectories is given in Fig. 1. The red dot represents an active individual; the leader is represented by the green dot.

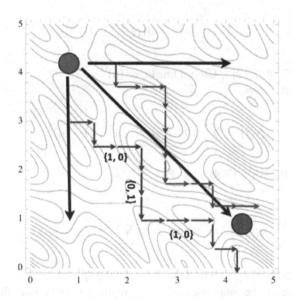

Fig. 1. The perturbation and the individual movement SOMA [10] (Color figure online)

3   Experiment Setup

In the experimental part, a *PRTVector* construction is altered to introduce the repulsive force. A pair of control parameters for perturbation (*PRT1* and *PRT2*) is introduced, following the pattern $PRT2 \geq PRT1 + 0.05$. During the construction of *PRTVector*, if random number $r < PRT1$, a value of -1 is assigned, if $r \geq PRT2$, the value of $+1$ is assigned. Otherwise, into the corresponding position in *PRTVector* is assigned a zero.

Using this mechanism, we conduct a tuning experiment for *PRT1* from 0 to 0.9 and *PRT2* from 0.05 to 1, by step 0.05 for both parameters, keeping the *PRT2* $\geq$ *PRT1* + 0.05 rule, leading to 209 combinations. For each setting, the SOMA algorithm is repeated 30 times, with random population initialization.

With accordance to general recommendations [1, 6, 7] and the focus of this study, the algorithm was set up as follows:

Pop. Size: 30; Migration loops: 100;
Step = 0.11; PathLength = 3;
D = 15;

The following set of three well-known benchmark functions (2)–(4) [19] was used for the experiments presented in this paper. We have selected a limited number of simpler functions due to the high number of possible parameter combinations and complex results analyses. Also, those simpler well-known functions increase the understandability of gained results and direct links into the behavior of the SOMA algorithm under different scenarios.

Rastrigin's function.

$$f_6(x) = \sum_{i=1}^{D} [x_i^2 - 10\cos(2\pi x_i) + 10] \tag{2}$$

Search Range: $[-5.12, 5.12]^D$; Glob. Opt. Pos.: $[0]^D$
Rosenbrock's function.

$$f_3(x) = \sum_{i=1}^{D-1} [100(x_i^2 - x_{i+1})^2 + (1 - x_i)^2] \tag{3}$$

Search Range: $[-10, 10]^D$; Glob. Opt. Pos.: $[0]^D$
Schwefel's function.

$$f_5(x) = 418.9829 \cdot D - \sum_{i=1}^{D} -x_i \sin(\sqrt{|x|}) \tag{4}$$

Search Range: $[-512, 511]^D$; Glob. Opt. Pos.: $[420.96]^D$
We investigate if the implementation of repulsive force into the perturbation operation in SOMA might improve the performance of the algorithm on such problems.

4 Results

As was outlined in the previous section, a tuning experiment the SOMA algorithm with non-binary PRTVector containing a repulsive mechanism was performed in this paper. A total number of 209 different scenarios (parameter settings) were tested.

4.1 Performance Visualization

In Figs. 2, 3 and 4, an overview of the results is given in the form of a 3D array plot, depicting the average CF value for 30 runs depending on PRT1 and PRT2 setting. Extremely high objective function values (CF) were clipped for clarity.

According to Fig. 2, an extreme increase of CF value is happening for values of PRT1 and PRT2 nearing their upper bound. In such cases, the repulsive mechanism is a dominant role (the number of generated -1 s is higher than the number of $+1$ s), and the individuals disperse. A similar trend is presented in Figs. 3 and 4.

For better clarity, we present a series of cross-views in Figs. 5, 6, 7, 8, 9, 10, 11, 12, 13 and 14 as an example or the case of Rastrigin function (the first scenario depicts a situation for original SOMA with no repulsively present). In all cases, it is clear, that a value of PRT2 over 0.8 leads to a sharp increase of the objective function value and therefore worse performance of the method.

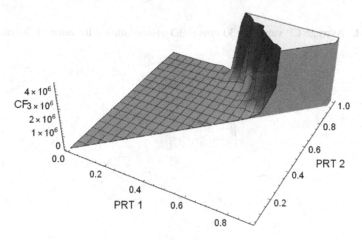

Fig. 2. Average CF value for 30 runs – 3D visualization – Rastrigin function

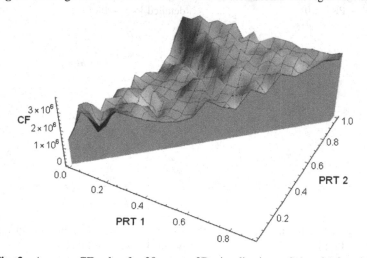

Fig. 3. Average CF value for 30 runs – 3D visualization – Schwefel function

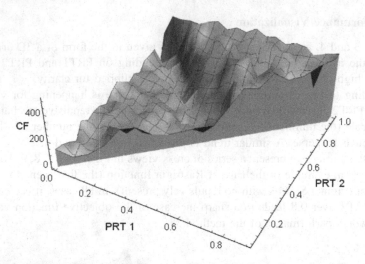

Fig. 4. Average CF value for 30 runs – 3D visualization – Rosenbrock function

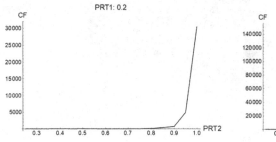

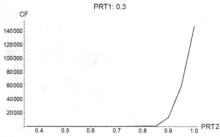

Fig. 5. Average CF value for 30 runs – 2D detailed look – PRT1: 0

Fig. 6. Average CF value for 30 runs – 2D detailed look – PRT1: 0

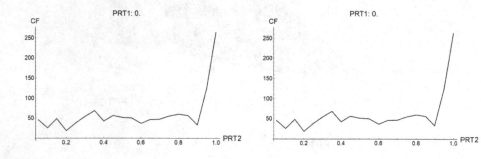

Fig. 7. Average CF value for 30 runs – 2D detailed look – PRT1: 0.2

Fig. 8. Average CF value for 30 runs – 2D detailed look – PRT1: 0.3

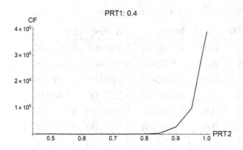

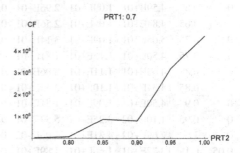

Fig. 9. Average CF value for 30 runs – 2D detailed look – PRT1: 0.4

Fig. 10. Average CF value for 30 runs – 2D detailed look – PRT1: 0.5

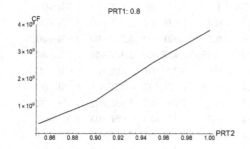

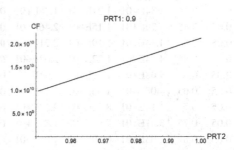

Fig. 11. Average CF value for 30 runs – 2D detailed look – PRT1: 0.6

Fig. 12. Average CF value for 30 runs – 2D detailed look – PRT1: 0.7

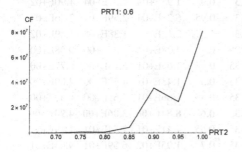

Fig. 13. Average CF value for 30 runs – 2D detailed look – PRT1: 0.8

Fig. 14. Average CF value for 30 runs – 2D detailed look – PRT1: 0.9

At first glance, it might seem that repulsive force implementation is not beneficial to SOMA; however evidence of opposite is presented in Table 1.

Table 1. Statistical overview of the results – Rastrigin function

PRT1	PRT2	Mean.	Std. dev.	Min	PRT1	PRT2	Mean.	Std. dev.	Min
0	0.05	3.70E+01	1.16E+01	1.95E+01	0.3	0.35	2.74E+01	1.11E+01	1.28E+01
0	0.1	3.77E+01	1.18E+01	2.22E+01	0.3	0.4	2.51E+01	8.05E+00	1.44E+01
0	0.15	4.28E+01	8.98E+00	2.39E+01	0.3	0.45	2.38E+01	9.89E+00	1.11E+01
0	0.2	3.71E+01	1.12E+01	1.90E+01	0.3	0.5	1.67E+01	5.77E+00	6.88E+00
0	0.25	3.63E+01	1.21E+01	1.55E+01	0.3	0.55	1.49E+01	5.53E+00	5.50E+00
0	0.3	3.99E+01	1.38E+01	2.22E+01	0.3	0.6	1.02E+01	3.35E+00	5.37E+00
0	0.35	3.64E+01	1.11E+01	1.76E+01	0.3	0.65	1.07E+01	7.17E+00	3.48E+00
0	0.4	4.16E+01	1.24E+01	1.40E+01	0.3	0.7	6.79E+01	2.46E+01	1.95E+01
0	0.45	4.75E+01	1.13E+01	2.60E+01	0.3	0.75	2.11E+02	5.55E+01	1.27E+02
0	0.5	4.94E+01	1.39E+01	2.62E+01	0.3	0.8	7.35E+02	3.34E+02	1.46E+02
0	0.55	4.65E+01	1.17E+01	2.70E+01	0.3	0.85	3.81E+03	1.97E+03	2.08E+02
0	0.6	4.59E+01	1.09E+01	2.96E+01	0.3	0.9	1.93E+04	1.07E+04	1.90E+02
0	0.65	4.84E+01	1.09E+01	2.56E+01	0.3	0.95	6.28E+04	2.59E+04	2.76E+04
0	0.7	5.05E+01	1.09E+01	3.14E+01	0.3	1	2.12E+05	1.38E+05	1.89E+02
0	0.75	4.50E+01	1.15E+01	1.81E+01	0.35	0.4	2.28E+01	7.90E+00	9.58E+00
0	0.8	4.63E+01	1.11E+01	3.06E+01	0.35	0.45	1.65E+01	6.18E+00	6.77E+00
0	0.85	4.34E+01	1.10E+01	2.33E+01	0.35	0.5	1.43E+01	4.72E+00	4.09E+00
0	0.9	4.50E+01	1.09E+01	1.51E+01	0.35	0.55	1.06E+01	4.65E+00	4.32E+00
0	0.95	1.10E+02	3.25E+01	5.52E+01	**0.35**	**0.6**	**8.84E+00**	**3.06E+00**	**4.94E+00**
0	1	2.69E+02	4.31E+01	1.96E+02	0.35	0.65	4.41E+01	2.54E+01	1.19E+01
0.05	0.1	3.72E+01	1.60E+01	1.89E+01	0.35	0.7	1.95E+02	6.59E+01	9.64E+01
0.05	0.15	3.96E+01	1.31E+01	1.51E+01	0.35	0.75	6.34E+02	2.94E+02	1.32E+02
0.05	0.2	3.47E+01	1.01E+01	1.80E+01	0.35	0.8	3.12E+03	1.04E+03	1.34E+03
0.05	0.25	3.73E+01	1.45E+01	1.80E+01	0.35	0.85	1.45E+04	7.11E+03	3.61E+03
0.05	0.3	3.37E+01	9.55E+00	1.67E+01	0.35	0.9	6.02E+04	4.79E+04	1.75E+02
0.05	0.35	4.01E+01	1.15E+01	1.79E+01	0.35	0.95	2.09E+05	2.28E+05	1.60E+02
0.05	0.4	3.84E+01	1.21E+01	1.78E+01	0.35	1	5.34E+05	3.81E+05	1.74E+02
0.05	0.45	4.29E+01	1.15E+01	2.46E+01	0.4	0.45	1.23E+01	5.08E+00	4.37E+00
0.05	0.5	4.34E+01	1.19E+01	2.21E+01	**0.4**	**0.5**	**8.90E+00**	**3.72E+00**	**3.83E+00**
0.05	0.55	4.44E+01	1.22E+01	2.08E+01	*0.4*	*0.55*	*7.43E+00*	*2.38E+00*	*2.17E+00*
0.05	0.6	4.08E+01	1.23E+01	1.91E+01	0.4	0.6	3.22E+01	1.80E+01	9.64E+00
0.05	0.65	4.08E+01	1.06E+01	2.13E+01	0.4	0.65	1.60E+02	4.47E+01	8.35E+01
0.05	0.7	4.14E+01	8.71E+00	2.22E+01	0.4	0.7	5.70E+02	2.83E+02	1.56E+02
0.05	0.75	3.31E+01	8.86E+00	1.54E+01	0.4	0.75	2.43E+03	2.06E+03	1.55E+02
0.05	0.8	2.95E+01	6.89E+00	1.85E+01	0.4	0.8	1.15E+04	5.06E+03	4.22E+03
0.05	0.85	3.06E+01	1.61E+01	8.26E+00	0.4	0.85	4.45E+04	2.18E+04	1.54E+04
0.05	0.9	6.09E+01	2.41E+01	1.56E+01	0.4	0.9	1.87E+05	9.02E+04	9.35E+04
0.05	0.95	1.22E+02	3.55E+01	5.40E+01	0.4	0.95	5.04E+05	3.17E+05	1.57E+02
0.05	1	8.92E+02	2.40E+02	5.83E+02	0.4	1	1.53E+06	8.34E+05	2.00E+02
0.1	0.15	3.46E+01	8.23E+00	1.60E+01	**0.45**	**0.5**	**7.72E+00**	**2.66E+00**	**2.00E+00**
0.1	0.2	3.36E+01	1.21E+01	1.40E+01	0.45	0.55	2.81E+01	1.49E+01	8.78E+00
0.1	0.25	3.49E+01	1.15E+01	1.61E+01	0.45	0.6	1.41E+02	6.27E+01	3.94E+01

(*continued*)

Table 1. (*continued*)

PRT1	PRT2	Mean.	Std. dev.	Min	PRT1	PRT2	Mean.	Std. dev.	Min
0.1	0.3	3.93E+01	1.16E+01	1.70E+01	0.45	0.65	4.00E+02	1.21E+02	1.35E+02
0.1	0.35	3.68E+01	1.27E+01	1.67E+01	0.45	0.7	2.12E+03	1.94E+03	1.46E+02
0.1	0.4	3.76E+01	1.22E+01	1.46E+01	0.45	0.75	1.00E+04	6.60E+03	2.04E+03
0.1	0.45	3.84E+01	9.36E+00	2.09E+01	0.45	0.8	5.61E+04	5.44E+04	1.80E+02
0.1	0.5	4.16E+01	1.33E+01	1.91E+01	0.45	0.85	1.58E+05	1.10E+05	1.46E+02
0.1	0.55	3.81E+01	1.19E+01	1.59E+01	0.45	0.9	5.01E+05	2.69E+05	2.24E+02
0.1	0.6	3.95E+01	1.15E+01	1.93E+01	0.45	0.95	1.63E+06	9.18E+05	2.20E+02
0.1	0.65	3.41E+01	8.46E+00	1.88E+01	0.45	1	4.16E+06	2.54E+06	1.60E+02
0.1	0.7	3.02E+01	8.72E+00	1.67E+01	0.5	0.55	1.14E+02	4.39E+01	5.61E+01
0.1	0.75	2.23E+01	6.08E+00	8.88E+00	0.5	0.6	3.67E+02	1.51E+02	1.29E+02
0.1	0.8	1.91E+01	9.26E+00	8.50E+00	0.5	0.65	1.28E+03	1.02E+03	1.69E+02
0.1	0.85	3.29E+01	1.42E+01	1.49E+01	0.5	0.7	8.97E+03	9.41E+03	1.68E+02
0.1	0.9	1.11E+02	2.69E+01	4.97E+01	0.5	0.75	3.73E+04	2.79E+04	1.46E+02
0.1	0.95	5.23E+02	1.64E+02	1.46E+02	0.5	0.8	1.40E+05	7.56E+04	1.88E+02
0.1	1	2.74E+03	1.22E+03	2.02E+02	0.5	0.85	5.13E+05	4.75E+05	1.49E+02
0.15	0.2	3.84E+01	1.08E+01	1.41E+01	0.5	0.9	1.56E+06	8.26E+05	2.09E+02
0.15	0.25	3.68E+01	1.20E+01	1.75E+01	0.5	0.95	4.88E+06	3.06E+06	2.15E+02
0.15	0.3	3.97E+01	1.43E+01	1.55E+01	0.5	1	1.29E+07	7.18E+06	2.25E+02
0.15	0.35	3.99E+01	1.26E+01	1.50E+01	0.55	0.6	8.94E+02	4.28E+02	1.90E+02
0.15	0.4	3.86E+01	1.04E+01	2.26E+01	0.55	0.65	6.62E+03	5.26E+03	1.53E+03
0.15	0.45	3.51E+01	8.03E+00	2.06E+01	0.55	0.7	3.51E+04	3.36E+04	1.57E+02
0.15	0.5	3.17E+01	1.00E+01	1.29E+01	0.55	0.75	1.31E+05	7.68E+04	1.63E+02
0.15	0.55	3.70E+01	9.70E+00	1.92E+01	0.55	0.8	4.50E+05	2.87E+05	1.96E+02
0.15	0.6	2.97E+01	8.15E+00	1.14E+01	0.55	0.85	2.00E+06	1.18E+06	3.66E+05
0.15	0.65	2.58E+01	7.13E+00	1.46E+01	0.55	0.9	4.10E+06	2.53E+06	1.90E+02
0.15	0.7	2.03E+01	5.28E+00	1.10E+01	0.55	0.95	1.53E+07	1.56E+07	2.12E+02
0.15	0.75	1.30E+01	4.05E+00	7.91E+00	0.55	1	3.33E+07	2.01E+07	1.65E+02
0.15	0.8	3.24E+01	2.03E+01	6.90E+00	0.6	0.65	3.34E+04	2.30E+04	7.87E+03
0.15	0.85	8.65E+01	2.79E+01	3.10E+01	0.6	0.7	1.10E+05	6.12E+04	2.17E+02
0.15	0.9	4.12E+02	9.09E+01	2.07E+02	0.6	0.75	4.66E+05	2.95E+05	1.61E+02
0.15	0.95	2.06E+03	6.14E+02	1.23E+03	0.6	0.8	1.47E+06	8.78E+05	2.01E+02
0.15	1	8.47E+03	3.44E+03	2.04E+02	0.6	0.85	4.77E+06	3.67E+06	1.59E+02
0.2	0.25	3.18E+01	8.92E+00	1.77E+01	0.6	0.9	1.28E+07	7.56E+06	3.54E+06
0.2	0.3	3.24E+01	8.65E+00	1.70E+01	0.6	0.95	4.37E+07	3.89E+07	2.06E+02
0.2	0.35	3.35E+01	1.02E+01	1.83E+01	0.6	1	1.03E+08	8.12E+07	1.87E+02
0.2	0.4	3.37E+01	1.02E+01	1.97E+01	0.65	0.7	3.53E+05	2.70E+05	5.12E+04
0.2	0.45	3.22E+01	9.01E+00	1.68E+01	0.65	0.75	1.28E+06	7.84E+05	2.00E+02
0.2	0.5	3.42E+01	9.10E+00	1.98E+01	0.65	0.8	3.89E+06	2.39E+06	2.14E+02
0.2	0.55	2.69E+01	7.50E+00	1.28E+01	0.65	0.85	1.38E+07	1.35E+07	1.62E+02
0.2	0.6	2.35E+01	6.71E+00	1.16E+01	0.65	0.9	3.44E+07	1.86E+07	2.15E+02
0.2	0.65	1.68E+01	4.43E+00	9.09E+00	0.65	0.95	1.21E+08	8.72E+07	1.68E+02
0.2	0.7	9.88E+00	4.35E+00	4.26E+00	0.65	1	2.78E+08	1.81E+08	8.81E+07

(continued)

Table 1. (*continued*)

PRT1	PRT2	Mean.	Std. dev.	Min	PRT1	PRT2	Mean.	Std. dev.	Min
0.2	0.75	1.99E+01	1.35E+01	7.25E+00	0.7	0.75	3.59E+06	3.32E+06	1.70E+02
0.2	0.8	8.55E+01	2.53E+01	4.56E+01	0.7	0.8	1.18E+07	1.06E+07	1.66E+02
0.2	0.85	3.50E+02	7.23E+01	2.52E+02	0.7	0.85	3.73E+07	2.75E+07	1.30E+02
0.2	0.9	1.29E+03	4.03E+02	7.04E+02	0.7	0.9	1.07E+08	5.99E+07	1.74E+02
0.2	0.95	5.76E+03	3.12E+03	1.65E+02	0.7	0.95	3.07E+08	1.73E+08	1.98E+02
0.2	1	2.22E+04	1.51E+04	1.55E+02	0.7	1	7.96E+08	4.70E+08	1.95E+02
0.25	0.3	3.34E+01	1.07E+01	2.03E+01	0.75	0.8	2.62E+07	2.30E+07	2.00E+02
0.25	0.35	3.24E+01	7.17E+00	1.60E+01	0.75	0.85	1.04E+08	8.05E+07	1.90E+02
0.25	0.4	2.99E+01	1.05E+01	1.65E+01	0.75	0.9	2.89E+08	1.80E+08	2.11E+02
0.25	0.45	2.88E+01	1.04E+01	1.26E+01	0.75	0.95	7.28E+08	4.28E+08	2.03E+02
0.25	0.5	2.54E+01	9.49E+00	1.03E+01	0.75	1	2.02E+09	1.38E+09	1.85E+02
0.25	0.55	2.12E+01	6.48E+00	9.93E+00	0.8	0.85	1.96E+08	1.22E+08	1.71E+02
0.25	0.6	1.53E+01	4.48E+00	7.62E+00	0.8	0.9	6.46E+08	4.68E+08	2.00E+02
0.25	**0.65**	**9.21E+00**	**3.49E+00**	**3.08E+00**	0.8	0.95	2.34E+09	1.46E+09	2.07E+02
0.25	0.7	1.22E+01	6.45E+00	4.40E+00	0.8	1	4.65E+09	2.77E+09	2.07E+02
0.25	0.75	7.17E+01	2.09E+01	2.65E+01	0.85	0.9	1.17E+09	8.48E+08	1.83E+02
0.25	0.8	2.94E+02	9.99E+01	1.67E+02	0.85	0.95	3.75E+09	1.78E+09	2.15E+02
0.25	0.85	1.09E+03	4.28E+02	1.46E+02	0.85	1	1.22E+10	6.85E+09	1.90E+02
0.25	0.9	4.82E+03	2.19E+03	2.11E+03	0.9	0.95	9.39E+09	4.72E+09	2.15E+02
0.25	0.95	1.96E+04	1.18E+04	6.93E+03	0.9	1	2.63E+10	1.36E+10	1.93E+02
0.25	1	7.93E+04	4.02E+04	1.78E+02					

4.2 Numerical Results

In the following Table 1, the full numerical results for Rastrigin function are presented. For each combination of PRT1 and PRT2, an average (mean) results are given, alongside with standard deviation, and minimal (best) result among the 30 runs.

The best mean result is given in bold number, the best five mean results (and corresponding parameters) are highlighted by grey color. It is clear that when compared with non-repulsive SOMA (PRT1 = 0), the performance of some repulsive variants is improved notably. Further, none of the best five results on Rastrigin function was achieved by SOMA algorithm without repulsivity.

Further, in Table 2, the best five results for every function are presented. According to a performed Friedman, rank test (alpha 0.05), the top five results do not differ with statistical significance.

As supplementary material, the full numerical results for all functions and the Friedman rank results visualization are provided at the A.I.Lab resource webpage (https://ailab.fai.utb.cz/resources/) and are not included in this paper for space limitation reasons.

4.3   Parameter Pool Visualization

In this section, we provide a parameter pool visualization for the top performing variants presented in Table 2. In Figs. 15, 16 and 17, the standard Box and Whiskers plot for the best performing setting of SOMA is presented. For each parameter (15 parameters, $D = 15$) the plot depicts the median, max, min and lower and upper quartiles (25% and 75%). The data are combined from the results of all 30 repeated runs.

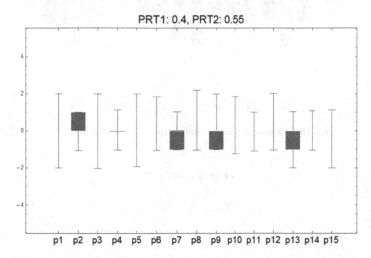

Fig. 15. Parameter pool visualization – Rastrigin function

Table 2. Statistical overview of the best five results

PRT1	PRT2	Mean	Std. dev.	Min	Max
Rastrigin					
0.25	0.65	9.21E+00	3.49E+00	3.08E+00	1.86E+01
0.35	0.6	8.84E+00	3.06E+00	4.94E+00	1.75E+01
0.4	0.5	8.90E+00	3.72E+00	3.83E+00	1.89E+01
0.4	**0.55**	**7.43E+00**	**2.38E+00**	**2.17E+00**	**1.11E+01**
0.45	0.5	7.72E+00	2.66E+00	2.00E+00	1.43E+01
Schwefel					
0	0.8	1.01E+06	3.05E+05	4.91E+05	1.92E+06
0	0.85	9.89E+05	2.38E+05	5.55E+05	1.49E+06
0	0.9	9.33E+05	2.08E+05	6.08E+05	1.38E+06
0	**0.95**	**9.36E+05**	**2.03E+05**	**5.76E+05**	**1.39E+06**
0.05	0.9	1.00E+06	2.15E+05	7.22E+05	1.45E+06
Rosenbrock					
0.25	0.75	4.73E+01	3.00E+01	4.87E+00	1.24E+02
0.3	**0.7**	**4.61E+01**	**3.24E+01**	**4.94E+00**	**1.46E+02**
0.4	0.6	5.25E+01	3.07E+01	1.64E+01	1.49E+02
0.45	0.5	5.47E+01	3.34E+01	5.20E+00	1.68E+02
0.45	0.55	5.38E+01	2.50E+01	1.71E+01	1.35E+02

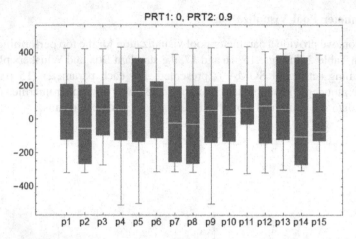

Fig. 16. Parameter pool visualization – Schwefel function

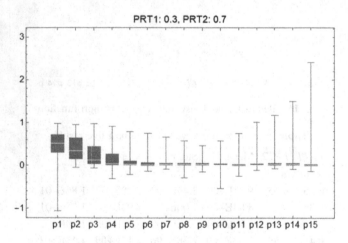

Fig. 17. Parameter pool visualization – Rosenbrock function

Finally, in Fig. 18, it is presented the *PRT1* and *PRT2* setting for the best five results for each function.

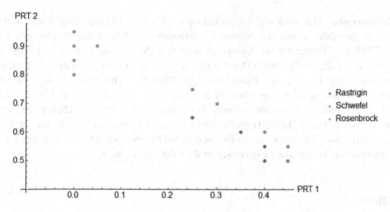

Fig. 18. The best five result - parameter setting

5   Conclusion

In this work, we present a detailed numerical evaluation and visualization of the performance and inner dynamic of SOMA algorithm with non-binary perturbation vector. In traditional variants of SOMA, the perturbation vector consists only from zeros and ones. We argue that implementation of repulsive force using third possible state (-1) in the perturbation vector might be beneficial for the algorithm. We have presented the computationally exhaustive tuning experiment and highlighted promising parameter settings.

The repulsive has been previously successfully implemented into other meta-heuristics, mostly swarm based. This study aims to show that such implementation into SOMA might also benefit its performance and encourage further research in this direction.

Based on the above presented experimental results, we conclude with the following discussion points, observations, and recommendations:

Introducing a repulsive mechanism into the SOMA algorithm seems likely to improve its performance on various problems, and further research in this direction is needed.

- The exact ratio between repulsive and attractive force probability in the perturbation is affected by the fitness landscape. However it seems that the approx. 40:50 ratio is promising.
- The repulsive variant did manage to outperform a nonrepulsive SOMA on two functions, but a more complex study is needed to support these encouraging initial findings.
- Given the computational expensiveness of this study, its scope is limited. We plan (as future research) to expand upon this study with more experiments using various

benchmark functions and dimensionality setting. Despite that, we hope that this study provides useful insight and encouragement for researches interested in SOMA and repulsivity mechanics in swarm-based methods in general.

Acknowledgements. This work was supported by the Ministry of Education, Youth and Sports of the Czech Republic within the National Sustainability Programme Project no. LO1303 (MSMT-7778/2014), further by the European Regional Development Fund under the Project CEBIA-Tech no. CZ.1.05/2.1.00/03.0089 and by Internal Grant Agency of Tomas Bata University under the Projects no. IGA/CebiaTech/2019/002. This work is also based upon support by COST (European Cooperation in Science & Technology) under Action CA15140, Improving Applicability of Nature-Inspired Optimisation by Joining Theory and Practice (ImAppNIO), and Action IC1406, High-Performance Modelling, and Simulation for Big Data Applications (cHiPSet). The work was further supported by resources of A.I.Lab at the Faculty of Applied Informatics, Tomas Bata University in Zlin (ailab.fai.utb.cz).

References

1. Eberhart, R., Kennedy, J.: Swarm Intelligence. The Morgan Kaufmann Series in Artificial Intelligence. Morgan Kaufmann, Burlington (2001)
2. Kennedy, J., Eberhart, R.: Particle swarm optimization. In: Proceedings of IEEE International Conference on Neural Networks, IV, pp. 1942–1948 (1995)
3. Dorigo, M.: Ant Colony Optimization and Swarm Intelligence. Springer, Heidelberg (2006). https://doi.org/10.1007/11839088
4. Parpinelli, R.S., Lopes, H.S.: New inspirations in swarm intelligence: a survey. Int. J. Bio-Inspired Comput. **3**(1), 1–16 (2011)
5. Zelinka, I., Jouni, L.: SOMA self-organizing migrating algorithm mendel. In: 6th International Conference on Soft Computing, Brno, Czech Republic (2000)
6. Zelinka, I.: SOMA - self organizing migrating algorithm. In: Babu, B.V., Onwubolu, G. (eds.) New Optimization Techniques in Engineering, vol. 33. Springer, Heidelberg (2004). https://doi.org/10.1007/978-3-540-39930-8_7. ISBN 3-540-20167X
7. Davendra, D., Zelinka, I.: Self-organizing Migrating Algorithm. Springer, Heidelberg (2016). https://doi.org/10.1007/978-3-319-28161-2
8. Nolle, L., et al.: Comparison of an self-organizing migration algorithm with simulated annealing and differential evolution for automated waveform tuning. Adv. Eng. Softw. **36** (10), 645–653 (2005)
9. Davendra, D., et al.: Discrete self-organising migrating algorithm for flow-shop scheduling with no-wait makespan. Math. Comput. Model. **57**(1–2), 100–110 (2013)
10. Kadlec, P., Raida, Z.: A novel multi-objective self-organizing migrating algorithm. Radioengineering **20**(4), 804–816 (2011)
11. Deep, K., Dipti: A self-organizing migrating genetic algorithm for constrained optimization. Appl. Math. Comput. 198(1), 237–250 (2008)
12. Deep, K., et al.: A new hybrid self organizing migrating genetic algorithm for function optimization. In: IEEE Congress on Evolutionary Computation, CEC 2007, pp. 2796–2803. IEEE (2007)
13. Bao, D.Q., Zelinka, I.: Obstacle avoidance for swarm robot based on self-organizing migrating algorithm. Procedia Comput. Sci. **150**, 425–432 (2019)

14. Sharma, S.K., Jain, Y.K.: Self organizing migration algorithm with curvelet based non local means method for the removal of different types of noise. Int. J. Comput. Sci. Inf. Secur. (IJCSIS) **16**(4), 320–330 (2018)
15. Fusek, R., Dobeš, P.: Pupil localization using self-organizing migrating algorithm. In: Zelinka, I., Brandstetter, P., Trong Dao, T., Hoang Duy, V., Kim, S.B. (eds.) AETA 2018. LNEE, vol. 554, pp. 207–216. Springer, Cham (2020). https://doi.org/10.1007/978-3-030-14907-9_21
16. Tomaszek, L., Lycka, P., Zelinka, I.: On the self-organizing migrating algorithm comparison by means of centrality measures. In: Zelinka, I., Brandstetter, P., Trong Dao, T., Hoang Duy, V., Kim, S.B. (eds.) AETA 2018. LNEE, vol. 554, pp. 335–343. Springer, Cham (2020). https://doi.org/10.1007/978-3-030-14907-9_33
17. Riget, J., Vesterstrøm, J.S.: A diversity-guided particle swarm optimizer - the ARPSO. Technical report 2, Department of Computer Science, University of Aarhus, Aarhus, Denmark (2002)
18. Pluhacek, M., Senkerik, R., Viktorin, A., Kadavy, T.: Particle swarm optimization with single particle repulsivity for multi-modal optimization. In: Rutkowski, L., Scherer, R., Korytkowski, M., Pedrycz, W., Tadeusiewicz, R., Zurada, J.M. (eds.) ICAISC 2018. LNCS (LNAI), vol. 10841, pp. 486–494. Springer, Cham (2018). https://doi.org/10.1007/978-3-319-91253-0_45
19. Dieterich, J.M., Hartke, B.: Empirical review of standard benchmark functions using evolutionary global optimization. arXiv preprint arXiv:1207.4318 (2012)

Boundary Strategies for Self-organizing Migrating Algorithm Analyzed Using CEC'17 Benchmark

Tomas Kadavy$^{(\boxtimes)}$ ⓘ, Michal Pluhacek ⓘ, Roman Senkerik ⓘ,
and Adam Viktorin ⓘ

Tomas Bata University in Zlin,
T.G. Masaryka 5555, 760 01 Zlin, Czech Republic
{kadavy,pluhacek,senkerik,aviktorin}@utb.cz

Abstract. This paper is focused on the influence of boundary strategies for the popular swarm-intelligence based optimization algorithm: Self-organizing Migrating Algorithm (SOMA). A similar extensive study was already performed for the most famous representative of swarm-based algorithm, which is Particle Swarm Optimization (PSO), and showed the importance of related research for other swarm-based techniques, like SOMA. The current CEC'17 benchmark suite is used for the performance comparison of the case studies, and the results are compared and tested for statistical significance using the Friedman Rank test.

Keywords: Self-organizing Migrating Algorithm · SOMA · Boundary · CEC17 · Friedman Rank test

1 Introduction

Almost every parameter, defined for an optimization task, is set within some boundary limits. These boundaries often exist as consequences to an optimized real-world problem (i.e., length of a screw must be only in positive numbers). The behavior of a metaheuristic optimization algorithm can lead to a trial solution that lies outside of the area of the feasible solution. In those cases, metaheuristic algorithm relies on some mechanism or method that address these violations and fixes the trial solution. Over the years, many different approaches were created. Some method can be universally used for almost every algorithm; others are strictly tuned for a specific problem or algorithm. The paper focuses on a few rather universal techniques and their influence on one selected metaheuristic optimization algorithm.

The research in the area of possible border methods and their influence on performance was already extensively carried out for another well-known optimization technique, which is the Particle Swarm Optimization (PSO) [1]. As the provided studies suggesting, it could be a challenging task [2, 3]. Similar research, to uncover the boundary methods influence, was also done for a Firefly Algorithm with exciting results [4].

© Springer Nature Switzerland AG 2020
A. Zamuda et al. (Eds.): SEMCCO 2019/FANCCO 2019, CCIS 1092, pp. 58–69, 2020.
https://doi.org/10.1007/978-3-030-37838-7_6

Furthermore, No Free Lunch (NFL) theorem [5] states that an algorithm which performs better on one type of optimization problems might be outperformed on a different kind of problems by another algorithm. This theorem also influences that one boundary method may not work well for the different optimization algorithm. Therefore, the development and testing of different boundary methods are crucial part of metaheuristic optimization algorithms tuning and effective set up.

The Self-Organizing Migrating Algorithm (SOMA) was initially developed in 1999 by Zelinka [6, 7]. SOMA takes inspiration in other swarm-based algorithms and also includes several different techniques. Namely, the discrete perturbation vector mimics the well-known mutation process that is used, for example, in evolution strategies and other classical evolutionary computing techniques. Another technique prepares SOMA for easy scalability due to the self-adaptation of movement over the search space. SOMA was already successfully used for continuous and discrete domains [8]. The possible universality of SOMA was used in the creation of its modification for solving multi-objective [9] or constrained optimization problems [10]. Like most of the other metaheuristic, the overall performance of SOMA depends on several user-defined parameters [11]. Again, as for almost other metaheuristic algorithms, the optimal settings of these parameters can vary over different optimization tasks.

This study is focused on as to how the canonical SOMA can handle the roaming individuals and what most common options exist and if they have any impact (and how significant) on the SOMA performance. Four relatively common borders methods are implemented and compared on well-known CEC'17 benchmark suite [12] and statistically evaluated using the Friedman Rank test [13]. The previous studies on PSO show the importance of the careful selection of the border method.

The paper is structured as follows. The canonical SOMA is described in the next Section. The implemented border methods are explained in Sect. 3. The experiment setting and results are presented in Sects. 4 and 5. Lastly, the conclusion of this study is presented in Sect. 6.

2 Self-organizing Migrating Algorithm

The main idea behind SOMA is based on the cooperation of individuals. Hence, an individual x is, in fact, also a solution to a defined optimization problem. The cooperation of these individuals should eventually lead to a global optimum. The so-called cooperation amongst individual is, by author, defined as a migration (1) of one particular individual from population towards another member of the whole population.

$$x_{i,j}^{k+1} = x_{i,j}^{k} + \left(x_{L,j}^{k} - x_{i,j}^{k} \right) \cdot t \cdot PRTVector_j \tag{1}$$

The $x_{i,j}^{k+1}$ represent a new position of an i-th individual in j-dimension for a next iteration step $k + 1$. Accordingly, the $x_{i,j}^{k}$ is a position of the same individual in k iteration. The $x_{L,j}^{k}$ is the position of a leader, which is selected based on the chosen SOMA strategy (SOMA strategies are described at the end of this section). Individual discrete steps between an i-th individual and selected leader $x_{L,j}^k$ are represented by t parameter. The best-found solution on this path is then transferred into a new generation. The t parameter is a collection of values starting from 0 to *Path* with increment (or step size) of *Step*. Both parameters *Path* and *Step* are user-defined; values for these parameters have a significant influence on the algorithm performance. However, some recommended values exist according to study [11], which are suitable for most optimization tasks.

The *PRTVector$_j$* mimics the mutation process and should be generated (2) for all the t steps. This vector determines in which dimensions j the i-th individual will migrate towards a leader and which dimensions stay unchanged. From the Eq. (2) is clear that the user-defined parameter *prt* has a direct impact on the resulting *PRTVector$_j$* and on the strength of a mutation during the migration. This *prt* parameter can be considered as a threshold value and is chosen in the range from 0 to 1 of a uniform distribution.

$$PRTVector_j = \begin{cases} \text{if } rand_j < prt, & 1 \\ \text{otherwise}, & 0 \end{cases} \tag{2}$$

Original SOMA describes several different strategies (most strategies affect only how the leader individual x_L is selected). For the purpose of this paper, two common types of strategies are described.

Strategy All-To-One. This easy to implement strategy will select for each migration cycle (one iteration of algorithm) one leader. The leader is selected based on its objective function value. All the remaining individual then migrate towards the leader. The pseudocode of this canonical SOMA is shown in Algorithm 1 SOMA – AllToOne.

Strategy All-To-All. The selection process of a leader is different for this strategy. One individual migrates towards all other individuals. After the end of the migration of a selected individual, this individual returns to its original position, and the process is repeated for the next individual. The migration cycle ends after all the individuals in population migrated towards each other, and all individuals then update their positions. This strategy shows stronger exploration abilities.

Algorithm 1 SOMA - AllToOne

```
 1: SOMA initialization
 2: while iteration < max_iteration do
 3: select leader x_L from population
 4: for i = 1 to NP do
 5:     for t = Step to Path do
 6:         generate PRTVector
 7:         migrate x_i to x_L
 8:     end for
 9:     save best x_i to new population
10:end for
11:record the best solution
12:end while
```

3 Boundary Methods

During each migration, the trial solution must be checked if it lies in the space of feasible solution (boundaries or range that is given to each dimension of a particular solution). If the solution, or rather a dimension of a solution, lies outside the defined boundaries, a specific correction has to be made. As being stated in Sect. 1, various approaches of how to handle infeasible solutions exist. Based on the similar research for PSO and FA algorithms [3, 4], these boundary methods may have a direct impact on the algorithm performance. Therefore, careful selection of the most suitable methods should be a priority for most applications. Also, if one process works well for one particular algorithm, there is no guarantee that the procedure will work for another optimization algorithm. For this paper, the most common methods were selected and compared together to show how they could affect the SOMA on different benchmark functions.

3.1 Clipping Method

For this method, the individual cannot cross the given boundaries in each dimension.

$$x'_{i,k} = \begin{cases} b^u_k, & \text{if } x_{i,k} > b^u_k \\ b^l_k, & \text{if } x_{i,k} < b^l_k \\ x_{i,k}, & \text{otherwise} \end{cases} \tag{3}$$

Where $x_{i,k}$ is the position of i-th individual in k-th dimension before boundary check, the $x'_{i,k}$ is a newly updated position after the boundary check and the b^u_k and b^l_k are the upper and lower boundary given to each dimension.

3.2 Random Method

If a particular dimension of an individual violates the given boundaries, the new value (position) for that dimension is newly generated between the lower b^l_k and upper b^u_k boundary (U stands for uniform distribution).

$$x'_{i,k} = \begin{cases} U\left(b^l_k, b^u_k\right), & \text{if } x_{i,k} > b^u_k \text{ OR } x_{i,k} < b^l_k \\ x_{i,k}, & \text{otherwise} \end{cases} \tag{4}$$

3.3 Reflection Method

The reflection method [2] resembles the behavior of a simple mirror. The violating dimension of an individual is reflected in the feasible space of solution.

$$x'_{i,k} = \begin{cases} b^u_k - \left(x_{i,k} - b^u_k\right), & \text{if } x_{i,k} > b^u_k \\ b^l_k + \left(b^l_k - x_{i,k}\right), & \text{if } x_{i,k} < b^l_k \\ x_{i,k}, & \text{otherwise} \end{cases} \tag{5}$$

3.4   Periodic Method

The last selected method takes advantages of an infinite space of solution. This endless space is achieved by mapping the individual back to the available area of solution using modulo function.

$$x'_{i,k} = b^l_k + \left(x_{i,k} \bmod \left(b^u_k - b^l_k\right)\right) \tag{6}$$

4   Experiment Setting

The benchmark set CEC'17 [12] was selected for the experiment. This benchmark includes 30 test functions divided into unimodal, multimodal, hybrid, and composite categories. However, the authors of the benchmark set recommend skipping the test function f_2 due to some technical difficulties. Only 10 and 30 dimension sizes were tested. The benchmark specifies the maximum number of function evaluations as $10 \cdot 000$ dim (dimension size). The benchmark also defines the boundary limits for all parameters as lower limit $b^l = -100$ and upper limit $b^u = 100$. Each test function was repeated for 51 independent runs, and the results were statistically evaluated.

The parameters of SOMA were set as $NP = 100$ (number of individuals), $prt = 0.3$, $Step = 0.11$ and $Path = 3$ according to the Authors [6, 7, 11].

The algorithm was programmed in C++ language (C++11) and executed on a PC with 64-bit Windows 10, AMD A8-7600 Radeon R7 3.1 GHz CPU and 4 GB RAM.

5   Results

In this section are presented the results of the performed experiments for both SOMA strategies (All-To-One and All-To-All). Firstly, the results overviews and comparisons are shown in Tables 2, 3, 4 and 5, which contain the simple statistic mean and standard deviation values.

The statistical significance was computed by the Friedman Rank test [13]. The null hypothesis that the mean is equal is rejected at the 5% level based on the Friedman

Table 1. P-values of Friedman Rank tests

Dimension size/strategy	All-To-One	All-To-All
dim = 10	7.41E−09	4.60E−04
dim = 30	4.22E−08	5.36E−10

Table 2. Statistical results for dimension 10, strategy All-To-One (mean, std. dev.)

f	Clipping		Random		Reflection		Periodic	
1	6.16E+09	3.52E+09	5.10E+08	3.17E+08	2.30E+09	1.30E+09	1.75E+09	1.16E+09
3	1.15E+04	6.40E+03	5.97E+03	3.17E+03	7.25E+03	4.23E+03	6.70E+03	3.85E+03
4	4.08E+02	1.17E+01	4.06E+02	2.64E+00	4.06E+02	4.81E+00	4.08E+02	1.10E+01
5	5.07E+02	3.18E+00	5.07E+02	2.26E+00	5.07E+02	2.54E+00	5.07E+02	2.04E+00
6	6.00E+02	6.75E−02	6.00E+02	8.28E−02	6.00E+02	7.11E−02	6.00E+02	7.63E−02
7	7.19E+02	2.83E+00	7.18E+02	3.08E+00	7.20E+02	3.42E+00	7.19E+02	3.14E+00
8	8.07E+02	2.33E+00	8.08E+02	3.14E+00	8.07E+02	2.34E+00	8.07E+02	2.79E+00
9	9.00E+02	2.17E−01	9.00E+02	1.66E−01	9.00E+02	3.24E−01	9.00E+02	1.88E−01
10	1.29E+03	1.38E+02	1.28E+03	1.50E+02	1.27E+03	1.63E+02	1.24E+03	1.41E+02
11	1.17E+03	2.33E+02	1.12E+03	3.23E+01	1.15E+03	1.87E+02	1.13E+03	7.00E+01
12	4.99E+07	5.72E+07	3.92E+06	2.66E+06	8.23E+06	6.08E+06	6.42E+06	5.43E+06
13	3.04E+05	5.29E+05	1.96E+04	2.87E+04	5.47E+04	8.03E+04	5.24E+04	8.68E+04
14	2.71E+03	2.13E+03	1.79E+03	6.83E+02	2.07E+03	1.28E+03	2.01E+03	1.25E+03
15	5.60E+03	5.09E+03	2.04E+03	7.25E+02	3.56E+03	2.46E+03	3.82E+03	2.80E+03
16	1.62E+03	3.88E+01	1.61E+03	2.47E+01	1.61E+03	2.32E+01	1.61E+03	3.27E+01
17	1.71E+03	9.10E+00	1.70E+03	3.25E+00	1.70E+03	4.93E+00	1.70E+03	5.64E+00
18	5.74E+05	1.72E+06	1.43E+04	1.01E+04	6.88E+04	1.19E+05	4.59E+04	4.74E+04
19	2.66E+04	3.91E+04	3.39E+03	2.01E+03	6.33E+03	9.19E+03	8.17E+03	1.05E+04
20	2.01E+03	1.76E+01	2.00E+03	6.12E−01	2.00E+03	4.82E−01	2.00E+03	3.04E+00
21	2.25E+03	5.59E+01	2.21E+03	1.47E+01	2.23E+03	3.89E+01	2.21E+03	2.13E+01
22	2.30E+03	2.35E+01	2.29E+03	3.06E+01	2.29E+03	2.57E+01	2.29E+03	2.96E+01
23	2.61E+03	3.62E+00	2.61E+03	9.87E+00	2.61E+03	3.59E+00	2.61E+03	6.32E+00
24	2.71E+03	8.16E+01	2.56E+03	4.89E+01	2.65E+03	1.00E+02	2.58E+03	8.25E+01
25	2.93E+03	2.11E+01	2.92E+03	2.10E+01	2.92E+03	2.05E+01	2.92E+03	2.73E+01
26	2.92E+03	7.97E+01	2.88E+03	9.23E+01	2.91E+03	5.02E+01	2.91E+03	7.25E+01
27	3.09E+03	1.98E+00	3.09E+03	2.83E+00	3.09E+03	2.83E+00	3.10E+03	2.75E+00
28	3.24E+03	1.16E+02	3.13E+03	5.13E+01	3.26E+03	6.97E+01	3.14E+03	4.39E+01
29	3.16E+03	1.25E+01	3.17E+03	1.06E+01	3.16E+03	1.26E+01	3.17E+03	1.18E+01
30	2.19E+06	3.28E+06	2.78E+05	4.01E+05	3.70E+05	6.91E+05	5.02E+05	6.47E+05

Table 3. Statistical results for dimension 10, strategy All-To-All (mean, std. dev.)

f	Clipping		Random		Reflection		Periodic	
1	9.73E+09	4.16E+09	7.64E+09	3.91E+09	8.75E+09	3.71E+09	8.69E+09	3.21E+09
3	1.51E+04	5.23E+03	1.48E+04	4.90E+03	1.53E+04	5.09E+03	1.36E+04	5.05E+03
4	9.43E+02	2.65E+02	1.04E+03	4.31E+02	9.77E+02	2.80E+02	1.11E+03	3.69E+02
5	5.87E+02	1.69E+01	5.92E+02	1.61E+01	5.86E+02	1.71E+01	5.83E+02	1.56E+01
6	6.55E+02	1.20E+01	6.54E+02	9.81E+00	6.53E+02	9.41E+00	6.53E+02	9.25E+00
7	9.70E+02	7.10E+01	9.45E+02	7.00E+01	9.58E+02	6.08E+01	9.73E+02	6.09E+01

(*continued*)

Table 3. (*continued*)

f	Clipping		Random		Reflection		Periodic	
8	8.79E+02	1.23E+01	8.76E+02	1.28E+01	8.75E+02	1.41E+01	8.78E+02	1.29E+01
9	2.31E+03	4.54E+02	2.12E+03	3.34E+02	2.19E+03	3.97E+02	2.18E+03	4.70E+02
10	2.61E+03	1.80E+02	2.56E+03	2.07E+02	2.50E+03	2.07E+02	2.51E+03	2.06E+02
11	2.05E+03	6.11E+02	2.06E+03	7.83E+02	2.09E+03	1.10E+03	2.01E+03	1.26E+03
12	2.40E+08	2.37E+08	1.90E+08	1.43E+08	2.46E+08	2.60E+08	2.13E+08	1.93E+08
13	1.24E+06	1.66E+06	1.18E+06	2.57E+06	1.55E+06	3.92E+06	8.88E+05	2.75E+06
14	2.17E+03	7.55E+02	2.00E+03	6.15E+02	1.99E+03	3.64E+02	1.94E+03	4.71E+02
15	7.52E+03	5.09E+03	5.86E+03	3.09E+03	6.43E+03	4.31E+03	6.07E+03	3.38E+03
16	2.06E+03	1.63E+02	2.01E+03	1.30E+02	2.01E+03	1.30E+02	2.00E+03	1.41E+02
17	1.92E+03	9.35E+01	1.90E+03	5.42E+01	1.89E+03	6.97E+01	1.90E+03	5.95E+01
18	4.68E+06	6.90E+06	5.52E+06	1.84E+07	1.79E+06	1.62E+06	2.25E+06	3.52E+06
19	1.52E+04	1.34E+04	1.29E+04	1.87E+04	1.33E+04	2.05E+04	1.18E+04	1.18E+04
20	2.22E+03	7.19E+01	2.21E+03	7.20E+01	2.22E+03	7.15E+01	2.19E+03	7.90E+01
21	2.30E+03	5.30E+01	2.29E+03	4.55E+01	2.30E+03	4.31E+01	2.28E+03	4.60E+01
22	2.97E+03	3.03E+02	2.99E+03	3.06E+02	2.98E+03	3.04E+02	2.91E+03	2.78E+02
23	2.71E+03	2.62E+01	2.71E+03	3.95E+01	2.72E+03	2.99E+01	2.72E+03	2.76E+01
24	2.83E+03	6.11E+01	2.82E+03	5.21E+01	2.80E+03	6.40E+01	2.82E+03	5.70E+01
25	3.44E+03	2.74E+02	3.30E+03	1.79E+02	3.44E+03	3.02E+02	3.38E+03	2.32E+02
26	3.85E+03	3.35E+02	3.76E+03	3.22E+02	3.86E+03	3.35E+02	3.73E+03	3.04E+02
27	3.19E+03	3.41E+01	3.19E+03	3.39E+01	3.19E+03	2.80E+01	3.20E+03	2.88E+01
28	3.63E+03	1.32E+02	3.59E+03	1.40E+02	3.59E+03	1.31E+02	3.57E+03	1.52E+02
29	3.37E+03	6.91E+01	3.35E+03	6.04E+01	3.34E+03	6.13E+01	3.35E+03	6.34E+01
30	6.17E+06	4.38E+06	6.41E+06	4.47E+06	5.15E+06	3.81E+06	6.98E+06	4.90E+06

Table 4. Statistical results for dimension 30, strategy All-To-One (mean, std. dev.)

f	Clipping		Random		Reflection		Periodic	
1	6.54E+10	1.26E+10	2.47E+10	6.11E+09	4.96E+10	9.39E+09	3.55E+10	8.00E+09
3	1.22E+05	2.31E+04	9.03E+04	1.99E+04	1.05E+05	1.96E+04	9.19E+04	2.13E+04
4	5.21E+02	3.05E+01	5.12E+02	2.18E+01	5.15E+02	2.20E+01	5.17E+02	2.50E+01
5	5.66E+02	2.02E+01	5.62E+02	2.20E+01	5.65E+02	2.21E+01	5.69E+02	2.60E+01
6	6.02E+02	8.55E−01	6.02E+02	1.10E+00	6.02E+02	6.57E−01	6.02E+02	9.74E−01
7	8.24E+02	2.81E+01	8.19E+02	2.66E+01	8.28E+02	2.82E+01	8.22E+02	3.03E+01
8	8.65E+02	2.25E+01	8.64E+02	2.53E+01	8.63E+02	2.38E+01	8.62E+02	2.00E+01
9	1.07E+03	1.99E+02	9.72E+02	7.41E+01	1.01E+03	1.12E+02	9.74E+02	6.70E+01
10	5.80E+03	8.76E+02	6.07E+03	7.19E+02	5.90E+03	9.09E+02	6.09E+03	8.92E+02
11	5.23E+03	4.66E+03	2.11E+03	1.14E+03	3.09E+03	2.45E+03	2.74E+03	1.66E+03
12	1.01E+10	3.68E+09	1.32E+09	6.59E+08	4.77E+09	1.83E+09	3.37E+09	1.35E+09
13	4.88E+09	3.28E+09	4.08E+08	4.25E+08	2.30E+09	1.56E+09	1.71E+09	1.80E+09

(*continued*)

Table 4. (*continued*)

f	Clipping		Random		Reflection		Periodic	
14	7.17E+05	1.60E+06	1.03E+05	1.02E+05	1.82E+05	2.72E+05	1.84E+05	2.67E+05
15	2.87E+08	2.96E+08	7.64E+06	1.42E+07	6.53E+07	9.49E+07	4.51E+07	6.71E+07
16	2.28E+03	2.00E+02	2.30E+03	2.06E+02	2.29E+03	2.64E+02	2.27E+03	2.26E+02
17	1.90E+03	1.20E+02	1.85E+03	8.57E+01	1.84E+03	9.37E+01	1.83E+03	6.93E+01
18	5.48E+06	7.60E+06	1.01E+06	6.42E+05	1.61E+06	1.64E+06	1.28E+06	1.31E+06
19	5.05E+08	4.40E+08	1.43E+07	1.70E+07	1.17E+08	1.45E+08	6.86E+07	7.13E+07
20	2.21E+03	1.28E+02	2.17E+03	8.82E+01	2.18E+03	1.03E+02	2.17E+03	8.49E+01
21	2.38E+03	2.86E+01	2.37E+03	2.47E+01	2.38E+03	2.51E+01	2.37E+03	2.17E+01
22	4.70E+03	2.66E+03	2.31E+03	4.88E+00	2.73E+03	1.28E+03	2.42E+03	7.33E+02
23	2.72E+03	1.59E+01	2.71E+03	1.79E+01	2.72E+03	1.91E+01	2.71E+03	1.88E+01
24	2.96E+03	2.98E+01	2.95E+03	3.77E+01	2.96E+03	3.42E+01	2.93E+03	3.94E+01
25	2.92E+03	2.29E+01	2.91E+03	1.91E+01	2.92E+03	2.37E+01	2.92E+03	2.01E+01
26	4.22E+03	4.48E+02	4.25E+03	3.36E+02	4.28E+03	3.78E+02	4.30E+03	3.06E+02
27	3.22E+03	1.05E+01	3.22E+03	1.08E+01	3.21E+03	1.40E+01	3.23E+03	1.12E+01
28	3.28E+03	3.01E+01	3.27E+03	2.78E+01	3.29E+03	1.69E+01	3.28E+03	2.82E+01
29	3.56E+03	2.72E+02	3.58E+03	2.28E+02	3.55E+03	2.14E+02	3.51E+03	1.17E+02
30	4.36E+08	3.83E+08	2.00E+07	1.48E+07	1.30E+08	1.20E+08	6.28E+07	4.93E+07

Table 5. Statistical results for dimension 30, strategy All-To-All (mean, std. dev.)

f	Clipping		Random		Reflection		Periodic	
1	6.40E+10	1.31E+10	6.24E+10	1.37E+10	6.25E+10	9.93E+09	5.86E+10	1.01E+10
3	1.24E+05	1.69E+04	1.14E+05	1.63E+04	1.19E+05	1.69E+04	1.10E+05	1.58E+04
4	1.56E+04	4.98E+03	1.52E+04	4.11E+03	1.54E+04	4.51E+03	1.43E+04	3.97E+03
5	9.53E+02	3.36E+01	9.39E+02	3.90E+01	9.31E+02	3.33E+01	9.34E+02	3.89E+01
6	6.90E+02	7.98E+00	6.84E+02	8.74E+00	6.86E+02	9.38E+00	6.87E+02	8.26E+00
7	2.26E+03	2.23E+02	2.27E+03	2.22E+02	2.32E+03	2.07E+02	2.21E+03	2.02E+02
8	1.21E+03	3.22E+01	1.20E+03	2.56E+01	1.19E+03	3.08E+01	1.19E+03	3.05E+01
9	1.52E+04	2.38E+03	1.51E+04	2.47E+03	1.53E+04	2.16E+03	1.52E+04	2.50E+03
10	8.21E+03	3.28E+02	8.14E+03	2.85E+02	8.08E+03	3.16E+02	8.17E+03	2.75E+02
11	1.01E+04	2.58E+03	8.75E+03	2.30E+03	8.61E+03	2.03E+03	8.79E+03	2.53E+03
12	9.24E+09	2.56E+09	8.05E+09	2.52E+09	8.53E+09	2.88E+09	8.55E+09	3.04E+09
13	5.89E+09	2.24E+09	4.91E+09	2.95E+09	5.20E+09	2.07E+09	4.97E+09	3.11E+09
14	1.05E+06	5.59E+05	7.13E+05	3.53E+05	8.39E+05	4.94E+05	1.01E+06	6.21E+05
15	5.91E+08	5.02E+08	5.39E+08	3.21E+08	4.58E+08	3.58E+08	5.50E+08	4.15E+08
16	4.71E+03	4.03E+02	4.56E+03	3.56E+02	4.65E+03	3.97E+02	4.69E+03	4.18E+02
17	3.32E+03	4.13E+02	3.14E+03	2.93E+02	3.22E+03	3.44E+02	3.17E+03	2.71E+02
18	1.24E+07	7.54E+06	9.49E+06	5.24E+06	1.15E+07	7.28E+06	1.05E+07	5.83E+06
19	6.81E+08	4.65E+08	5.85E+08	4.92E+08	5.64E+08	3.76E+08	5.59E+08	3.50E+08

(*continued*)

Table 5. (*continued*)

f	Clipping		Random		Reflection		Periodic	
20	2.91E+03	8.67E+01	2.85E+03	9.59E+01	2.88E+03	9.64E+01	2.85E+03	1.03E+02
21	2.71E+03	3.16E+01	2.70E+03	3.05E+01	2.70E+03	3.92E+01	2.69E+03	2.75E+01
22	9.38E+03	5.27E+02	8.92E+03	1.06E+03	8.97E+03	1.02E+03	8.96E+03	6.38E+02
23	3.29E+03	8.79E+01	3.27E+03	6.98E+01	3.30E+03	7.05E+01	3.30E+03	8.40E+01
24	3.48E+03	9.66E+01	3.49E+03	1.03E+02	3.54E+03	9.84E+01	3.51E+03	8.58E+01
25	7.95E+03	1.29E+03	7.34E+03	1.30E+03	7.70E+03	1.36E+03	7.44E+03	1.16E+03
26	1.05E+04	8.74E+02	1.03E+04	8.95E+02	1.04E+04	8.54E+02	1.04E+04	8.36E+02
27	3.85E+03	1.43E+02	3.85E+03	1.40E+02	3.77E+03	1.70E+02	3.90E+03	1.39E+02
28	7.47E+03	1.04E+03	7.16E+03	8.10E+02	6.93E+03	8.16E+02	7.04E+03	9.98E+02
29	6.05E+03	4.95E+02	5.76E+03	4.48E+02	5.90E+03	5.22E+02	5.93E+03	4.42E+02
30	6.69E+08	3.93E+08	4.88E+08	3.14E+08	4.81E+08	2.54E+08	4.49E+08	2.54E+08

Rank test. The corresponding p-values of Friedman Rank test are presented in Table 1. If the p-value is lower than 0.05, the further Friedman rankings are relevant.

In Fig. 1, the results of Friedman ranking are shown for combinations of both SOMA strategies, and both tested dimension sizes 10 and 30. The lower the rank is, the better is the performance of the labeled strategy. The Nemenyi Critical Distance post-hoc test for multiple comparisons was used to compute the critical distance for each Friedman rank set. The critical distance is represented as a dashed line from the best-ranked boundary method. The critical distance (CD) value for this experiment has been calculated as 0.656757; according to the definition given in (7) and value $q_a = 2.56892$; using $k = 4$ boundary methods and a number of data sets $N = 51$ (51 repeated runs).

$$CD = q_a \sqrt{k(k+1)/(6N)} \tag{7}$$

Based on the obtained ranks for All-To-One strategy, Random and Periodic methods are the most promising ones. For All-To-All strategy, the situation is similar with one exception that even the Reflection method seems to be also promising to use. From the ranks, it is clear that the worst option is the Clipping method.

Two distinct groups of methods can be observed. The first group (Clipping, Reflection, and Periodic) keep the consistent migration of individual over the search space without a loss of information from the previous positions. The second group, including only one Random method, more resembling the stochastic search optimization [14]. This method may help to SOMA escape some traps of a local minimum. This behavior seems to be the best work solution for both tested SOMA strategies. Although the remaining Periodic method has a slightly worst ranks, the individuals are keeping their movement directions and search patterns; thus, this method may be an exciting choice for some modern SOMA modifications.

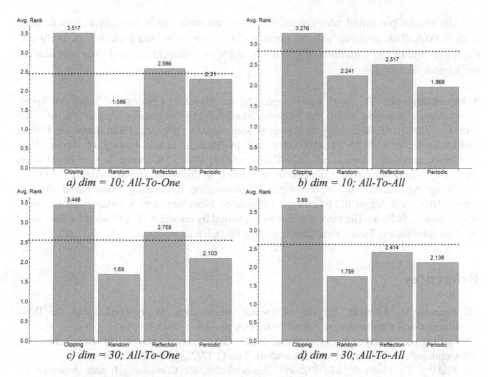

Fig. 1. Friedman Rank tests for selected border methods

6 Conclusion

In this study, the influence of four selected border methods on the performance of canonical SOMA is tested and evaluated. Due to the increasing complexity of optimization problems, it is essential to focus not only on the development of new algorithms but also on deeper insights, analyses and improvement of existing and effective algorithms. The new benchmark set CEC 2017 was selected for the performance comparison because it should represent the most recent collection of artificial optimization problems. Based on the obtained results, it may be concluded that the Random and Periodic methods are the most promising ones and that the Random method is probably introducing more stochastic behavior to SOMA.

On the other hand, the Periodic method is keeping the movement directions and search patterns of individuals, thus resulting in a more natural and predictable response according to the original population dynamics model. Based on the obtained results, it is clear that the worst method for canonical SOMA, both for All-To-One and All-To-All strategies, is Clipping method. This may be caused by the fact that in some cases, the individuals have tendencies to move to the next trial solution, but it is forced to stay on the fixed borders of feasible space—all of the observations suit for both dimension settings with almost the same results.

The results presented here are useful as an empirical study for researchers dealing with SOMA. This research will continue in the future on related topic lies in analyzing the frequency of the violations of given boundaries by individuals and their spread over the hyperspace of solution.

Acknowledgments. This work was supported by the Ministry of Education, Youth and Sports of the Czech Republic within the National Sustainability Programme Project no. LO1303 (MSMT-7778/2014), further by the European Regional Development Fund under the Project CEBIA-Tech no. CZ.1.05/2.1.00/03.0089 and by Internal Grant Agency of Tomas Bata University under the Projects no. IGA/CebiaTech/2019/002. This work is also based upon support by COST (European Cooperation in Science & Technology) under Action CA15140, Improving Applicability of Nature-Inspired Optimisation by Joining Theory and Practice (ImAppNIO), and Action IC1406, High-Performance Modelling, and Simulation for Big Data Applications (cHiPSet). The work was further supported by resources of A.I.Lab at the Faculty of Applied Informatics, Tomas Bata University in Zlin (ailab.fai.utb.cz).

References

1. Kennedy, J., Eberhart, R.: Particle swarm optimization. In: Proceedings of the IEEE International Conference on Neural Networks, pp. 1942–1948 (1995)
2. Helwig, S., Branke, J., Mostaghim, S.M.: Experimental analysis of bound handling techniques in particle swarm optimization. TEVC **17**(2), 259–271 (2013)
3. Kadavy, T., Pluhacek, M., Viktorin, A., Senkerik, R.: Comparing border strategies for roaming particles on single and multi-swarm PSO. In: Silhavy, R., Senkerik, R., Kominkova Oplatkova, Z., Prokopova, Z., Silhavy, P. (eds.) CSOC 2017. AISC, vol. 573, pp. 528–536. Springer, Cham (2017). https://doi.org/10.1007/978-3-319-57261-1_52
4. Kadavy, T., Pluhacek, M., Viktorin, A., Senkerik, R.: Boundary strategies for firefly algorithm analysed using CEC17 benchmark. In: Proceedings of European Council for Modelling and Simulation, ECMS (2018)
5. Wolpert, D.H., Macready, W.G.: No free lunch theorems for optimization. IEEE Trans. Evol. Comput. **1**(1), 67–82 (1997)
6. Zelinka, I.: SOMA—self-organizing migrating algorithm. In: Onwubolu, G.C., Babu, B.V. (eds.) New Optimization Techniques in Engineering. Studies in Fuzziness and Soft Computing, vol. 141, pp. 167–217. Springer, Heidelberg (2004). https://doi.org/10.1007/978-3-540-39930-8_7
7. Zelinka, I.: SOMA—self-organizing migrating algorithm. In: Davendra, D., Zelinka, I. (eds.) Self-organizing Migrating Algorithm. Studies in Computational Intelligence, vol. 626, pp. 3–49. Springer, Cham (2016). https://doi.org/10.1007/978-3-319-28161-2_1
8. Davendra, D., Zelinka, I., Pluhacek, M., Senkerik, R.: DSOMA—discrete self organising migrating algorithm. In: Davendra, D., Zelinka, I. (eds.) Self-organizing Migrating Algorithm. Studies in Computational Intelligence, vol. 626, pp. 51–63. Springer, Cham (2016). https://doi.org/10.1007/978-3-319-28161-2_2
9. Kadlec, P., Raida, Z.: Multi-objective self-organizing migrating algorithm. In: Davendra, D., Zelinka, I. (eds.) Self-organizing Migrating Algorithm. Studies in Computational Intelligence, vol. 626, pp. 83–103. Springer, Cham (2016). https://doi.org/10.1007/978-3-319-28161-2_4

10. Singh, D., Agrawal, S., Deep, K.: C-SOMAQI: self organizing migrating algorithm with quadratic interpolation crossover operator for constrained global optimization. In: Davendra, D., Zelinka, I. (eds.) Self-organizing Migrating Algorithm. Studies in Computational Intelligence, vol. 626, pp. 147–165. Springer, Cham (2016). https://doi.org/10.1007/978-3-319-28161-2_7

11. Čičková, Z., Lukáčik, M.: Setting of control parameters of SOMA on the base of statistics. In: Davendra, D., Zelinka, I. (eds.) Self-organizing Migrating Algorithm. Studies in Computational Intelligence, vol. 626, pp. 255–275. Springer, Cham (2016). https://doi.org/10.1007/978-3-319-28161-2_12

12. Awad, N.H., et al.: Problem definitions and evaluation criteria for CEC 2017 special session and competition on single-objective real-parameter numerical optimization (2016)

13. Demsar, J.: Statistical comparisons of classifiers over multiple data sets. J. Mach. Learn. Res. 7(Jan), 1–30 (2006)

14. Bergstra, J., Bengio, Y.: Random search for hyper-parameter optimization. J. Mach. Learn. Res. 13(Feb), 281–305 (2012)

MOEA with Approximate Nondominated Sorting Based on Sum of Normalized Objectives

Vikas Palakonda and Rammohan Mallipeddi[(✉)]

School of Electronics Engineering, Kyungpook National University,
Daegu 702 701, South Korea
vikas11475@gmail.com, mallipeddi.ram@gmail.com

Abstract. Pareto based selection techniques are extensively implemented in the multi-objective evolutionary algorithms (MOEAs), to tackle the many-objective optimization problems (MaOPs). In Pareto-dominance based MOEAs (PDMOEAs), nondominated sorting (NDS) plays a prominent role in preserving the elite solutions during mating and environmental selection. Although, NDS is an inevitable procedure in the evolution of PDMOEAs, computational complexity issues enhances the difficulty to adopt NDS approaches. Various methodologies were suggested in literature to overcome complexity issues, but these approaches deteriorate drastically for higher objectives. Recently, an approximate efficient NDS, (AENS) is proposed that utilize three objective comparisons to establish the dominance relation. In this paper, we propose an improved version of AENS, in which maximum two objective comparisons are required to determine the dominance relation. To evaluate the performance of our algorithm, experiments are done on seven different test problems and the experiment results have proved the effectiveness of proposed method in improving the convergence of different MOEAs.

Keywords: Approximate Non-dominated Sorting · Sum of normalized objectives · Convergence · Diversity

1 Introduction

Multi-objective optimization deals with multiple conflicting objectives that ought to be solved simultaneously and the problems associated with multiple objectives are termed as multi-objective optimization problems (MOPs). While handling MOPs, a set of nondominated solutions, termed as Pareto-optimal solutions are to be obtained instead of single optimum solution due to the conflicting nature of the objectives [1]. Evolutionary algorithms (EAs) became popular in solving the MOPs, as they have the ability to obtain Pareto-optimal solutions in one single run. The goal of MOEAs is to achieve proper trade-off between convergence and diversity among the obtained Pareto-optimal solutions which can be considered as a difficult task [2, 3].

To achieve set of Pareto-optimal solutions with balanced convergence and diversity, several selection strategies were proposed in the past, among which, the Pareto-dominance based approaches have proved to be effective as they assign more priority to

A. Zamuda et al. (Eds.): SEMCCO 2019/FANCCO 2019, CCIS 1092, pp. 70–78, 2020.
https://doi.org/10.1007/978-3-030-37838-7_7

the solutions with better Pareto rank. In other words, in PDMOEAs, the candidate solutions are ranked based on their dominance relationship and solutions that are nondominated are assigned high priority during mating and environmental selection process [2, 3]. In addition, among the solutions that have same rank, solutions that are well separated or diverse are given priority. However, the performance of PDMOEAs is effective in solving the problems with two and three objectives and gradually deteriorates as the objectives increases i.e., for MaOPs. The deterioration in the performance of the PDMOEAs is due to rapid increase in the ratio of nondominated solutions as the number of objectives increase. Therefore, PDMOEAs rely on the secondary selection criterion that promotes diversity but not convergence leading to performance degradation [3–5]. In literature, various approaches are proposed to overcome this problem but few works were dedicated to deal with the complexity issues of the NDS approach.

In PDMOEAs, to overcome the computational complexity associated with non-dominated sorting, various methods such as fast non-dominated sort [6], climbing sort and deductive sort [7], corner sort [8], efficient non-dominated sorting (ENS) [9] and nondominated sorting based on sum of objectives [10] have been proposed. These methods try to reduce the number of unnecessary objective comparisons during NDS procedure. However, in [11], an approximate efficient NDS approach (AENS) was proposed for PDMOEAs to solve MaOPs. In AENS, the initial population is sorted on the first objective function and the solutions are assigned to the fronts in the sorted order. In addition, to determine the dominance relationship between two solutions, a maximum of three objective comparisons are performed. In case, if dominance relation cannot be established within the three comparisons, the solution is assumed to be dominated by the solution that is already assigned to the fronts. Hence, with the reduced objective comparisons, PDMOEAs with AENS demonstrate better convergence characteristics [11]. The improved convergence can be attributed to the reduced number of nondominated solutions. In other words, AENS can segregate solutions effectively and helps prioritizing nondominated solutions with better convergence.

In this paper, we propose an improved approximate NDS based on sum of normalized objectives (ASNDS) with reduced complexity of AENS. In the proposed approximate NDS approach, the solutions are sorted based on sum of normalized objectives and at most two objective comparisons are required to determine the dominance relation between two solutions. Similar to AENS, if the dominance relation cannot be determined after two objective comparisons the solution in the fronts is assumed to be dominating the solution to be assigned. Hence, the proposed approximate NDS approach is computationally less expensive and improves the convergence at the cost of diversity.

The rest of the paper is organized as follows. In Sect. 2, we have explained the proposed methods in detail. In Sect. 3, experimental results are presented and the Sect. 4 concludes the paper.

2 Proposed Method

In this section, a detailed explanation about the proposed approximate NDS is presented. In the proposed algorithm, the candidate solutions are sorted based on the sum of the normalized objectives and the solutions are assigned to the fronts according to the sorted order. The main objective of sorting solutions based on sum of normalized objectives is that a particular solution cannot dominate a solution with less value in terms of sum of normalized objectives than that solution. However, the range-dependence of the solutions is removed with the help of normalization.

2.1 Proposed Approximate NDS Based on Sum of Objectives

In this section, we have presented briefly the proposed approximate NDS based on sum of normalized objectives. This approach mainly consists of the two steps. First, all the candidate solutions in the population are sorted in an ascending order according to sum of normalized objectives. Second, the proposed algorithm performs at most two objective comparisons between the solutions to determine the dominance relation. The comparisons are performed in the sorted order of the population and the solutions are assigned to the non-dominated fronts one by one, starting from the first solution to the last one. If the algorithm fails to determine the dominance relation among two solutions within the two objective comparisons, then the solution that has already been assigned to a non-dominated front will be considered to dominate the one to be assigned to the front. The dominated solution will be checked for nondomination in the next fronts and if it is dominated by the solutions in all the fronts than the corresponding solution is assigned to the new front. This procedure repeats until all solutions in the population are assigned to a front.

Normalization of objectives is done as mentioned below. Let X is a solution with M objectives and the normalized objective value of solution X in i^{th} objective is given as follows [12]

$$X_{norm_i} = \frac{(X_i - f_{min})}{(f_{max} - f_{min})}$$

where X_{norm_i} and X_i are the normalized and actual objective value of solution X in i^{th} objective, f_{min} and f_{max} are minimum and maximum values in ith objective.

The detailed description of the proposed approximate NDS is explained through the Figs. 1 and 2. Let us consider a 4-objective minimization problem to be optimized by an MOEA as shown in Fig. 1. At first, all the objectives are normalized and solutions are sorted based on sum of the normalized objectives as shown in Fig. 1. From the Fig. 1, we can observe that the solution S_4 is first solution in the sorted order, considered as the nondominated solution and assigned to the first front. To assign the solution S_2 to the nondominated front, it has to be compared with the solution preceding to solution S_2 in the sorted order, solution S_4. Form the Fig. 2(a), we can observe that both the solutions are compared based on two objectives, minimum objective value in solution S_2 and maximum objective value in solution S_4. As the solution S_4 is unable to dominate the solution S_2 in both the objective, the solution S_2 is considered as the non-dominated solution and assigned to the first nondominated front. From the Fig. 2(b), comparison between the solutions S_4 and S_8 are presented and by comparing them on both the objectives, we can observe that solution S_4 dominates solution S_8 on both objectives.

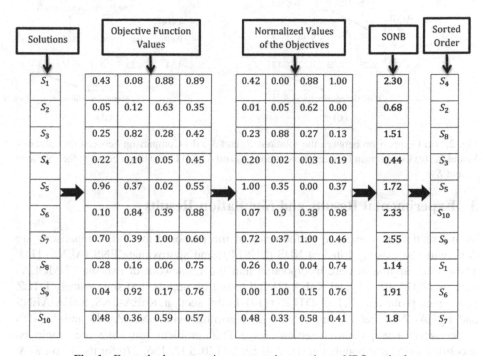

Solutions	Objective Function Values				Normalized Values of the Objectives				SONB	Sorted Order
S_1	0.43	0.08	0.88	0.89	0.42	0.00	0.88	1.00	2.30	S_4
S_2	0.05	0.12	0.63	0.35	0.01	0.05	0.62	0.00	0.68	S_2
S_3	0.25	0.82	0.28	0.42	0.23	0.88	0.27	0.13	1.51	S_8
S_4	0.22	0.10	0.05	0.45	0.20	0.02	0.03	0.19	0.44	S_3
S_5	0.96	0.37	0.02	0.55	1.00	0.35	0.00	0.37	1.72	S_5
S_6	0.10	0.84	0.39	0.88	0.07	0.9	0.38	0.98	2.33	S_{10}
S_7	0.70	0.39	1.00	0.60	0.72	0.37	1.00	0.46	2.55	S_9
S_8	0.28	0.16	0.06	0.75	0.26	0.10	0.04	0.74	1.14	S_1
S_9	0.04	0.92	0.17	0.76	0.00	1.00	0.15	0.76	1.91	S_6
S_{10}	0.48	0.36	0.59	0.57	0.48	0.33	0.58	0.41	1.8	S_7

Fig. 1. Example demonstrating proposed approximate NDS method.

Therefore, solution S_8 is considered as dominated solution irrespective of other objectives and assigned to the next nondominated level, second front. In Fig. 2(c) & (d), the comparison between solution S_3 with solutions S_4 & S_2 respectively is presented. It can be noted that both the solutions are unable to dominate the solution S_3 and hence the solution S_3 is considered as nondominated solution and assigned to first front. The similar procedure is followed until all the solutions are assigned to the fronts.

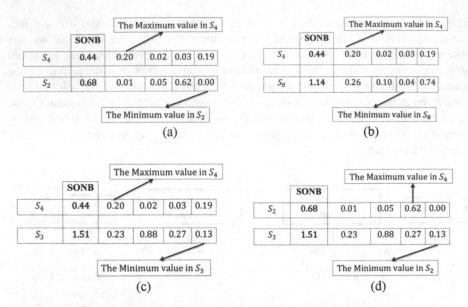

Fig. 2. (a) Comparison between the solutions S_4 and S_2; (b) Comparison between the solutions S_4 and S_8; (c) Comparison between the solutions S_4 and S_3; (d) Comparison between the solutions S_2 and S_3;

3   Experimental Results and Simulation Results

In this section, to depict the performance of the proposed method, comparisons are done with the existing Efficient NDS (ENS) [9] and approximate ENS (AENS) [11]. These three approaches are incorporated into the framework of existing PDMOEAs such as NSGA-III [13] and KnEA [2] and tested on popular test suite namely DTLZ benchmark problems (DTLZ1-DTLZ7) [14]. In this section, KnEA-ENS, KnEA-AENS and KnEA-SNDS refer to the KNEA [2] algorithm with ENS, AENS and ASNDS respectively. The parameter settings for DTLZ problem test suite are employed according to [5] and population size N is set as 120, 132, 156, 276 for the 4-, 6-, 8- & 10-objectives. The number of generations adopted in the experiments is 700 for DTLZ1 problem, 1000 for DTLZ3 problem and 250 for the remaining problems (DTLZ2 & DTLZ4-DTLZ7). All the algorithms considered for comparison are simulated for 30 runs and the final obtained populations is preserved for the comparison.

To demonstrate the effectiveness of proposed ASNDS, three performance indicators namely Hypervolume (HV) [15, 16], Spread [17] and Generational Distance (GD) [18] are used. HV indicator considers both the convergence and diversity whereas Spread and GD accounts for diversity and convergence respectively. Algorithm with higher HV and lower Spread and lower GD is regarded as the best performing approach. We have reported the mean and standard deviation values of the performance indicators such as Hypervolume, Spread and Generational distance in the Tables 1 and 2. For an accurate comparison, we have conducted significance test, highlighted the best performing approach in grey shade, and bold and the second best approach with grey shade.

Table 1. Mean and standard deviation values of hypervolume results of the proposed method with ENS and A_ENS by incorporating them in existing KNEA

Problem	M	Hypervolume			Spread			GD		
		KnEA			KnEA			KnEA		
		ENS	AENS	ASNDS	ENS	AENS	ASNDS	ENS	AENS	ASNDS
DTLZ1	4	0.6684	0.7418	0.5502	0.8435	0.9511	0.9634	0.0856	0.1017	0.0126
		0.1406	0.0780	0.0392	0.5759	0.6026	0.1993	0.1661	0.1954	0.0517
	6	0.5293	0.6317	0.4846	1.7288	1.1822	1.1519	0.6182	0.1478	0.0222
		0.0949	0.0928	0.0462	0.3632	0.4390	0.1703	0.5425	0.3012	0.1144
	8	0.5691	0.5720	0.6863	1.7526	1.0683	1.0399	1.1705	0.0400	0.0028
		0.3000	0.3743	0.2770	0.2564	0.2374	0.0634	1.0190	0.1483	0.0102
	10	0.8981	0.7282	0.8213	1.5294	1.0108	1.0008	2.7057	0.0311	0.0022
		0.1734	0.3429	0.2450	0.0695	0.0281	0.0009	0.6118	0.0920	0.0044
DTLZ2	4	0.5491	0.5463	0.5206	0.2970	0.4553	0.7781	0.0048	0.0044	0.0028
		0.0055	0.0060	0.0164	0.0797	0.0857	0.0842	0.0002	0.0003	0.0003
	6	0.7079	0.7107	0.6938	0.1914	0.4534	0.8033	0.0051	0.0044	0.0025
		0.0108	0.0105	0.0174	0.0933	0.0885	0.1077	0.0004	0.0005	0.0004
	8	0.9448	0.9436	0.9422	0.1483	0.4121	0.7082	0.0071	0.0046	0.0035
		0.0053	0.0026	0.0029	0.1172	0.0882	0.1050	0.0043	0.0003	0.0006
	10	0.9779	0.9989	0.9988	0.2379	0.4673	0.7651	0.0433	0.0054	0.0043
		0.0133	0.0001	0.0001	0.1057	0.0887	0.0820	0.0100	0.0005	0.0005
DTLZ3	4	0.4251	0.3804	0.2795	1.2991	1.2521	1.2069	0.4304	0.4478	0.0711
		0.0732	0.0722	0.0522	0.4752	0.4321	0.2395	0.6395	0.8026	0.2627
	6	0.9984	0.7938	1.0000	1.4710	1.0681	1.0029	6.7077	0.4479	0.0030
		0.0035	0.3115	0.0002	0.0716	0.1131	0.0008	4.0270	0.8582	0.0045
	8	0.6938	0.9877	0.9804	1.5158	1.0569	1.0249	7.8655	0.2093	0.1165
		0.4377	0.0501	0.1013	0.1739	0.1240	0.0534	3.4792	0.5088	0.3075
	10	1.0000	0.9995	0.9998	1.4835	1.0146	1.0024	3.6133	0.0639	0.0216
		0.0000	0.0006	0.0005	0.0734	0.0207	0.0059	1.6086	0.0943	0.0533
DTLZ4	4	0.5522	0.5471	0.5348	0.3015	0.3857	0.7710	0.0050	0.0046	0.0030
		0.0053	0.0256	0.0274	0.1050	0.0730	0.0951	0.0003	0.0003	0.0002
	6	0.7423	0.7478	0.7358	0.1961	0.3713	0.7300	0.0054	0.0043	0.0028
		0.0102	0.0070	0.0148	0.1325	0.1171	0.0930	0.0017	0.0004	0.0004
	8	0.9819	0.9819	0.9823	0.1939	0.3487	0.7027	0.0077	0.0045	0.0028
		0.0015	0.0009	0.0008	0.1597	0.1192	0.0872	0.0035	0.0004	0.0005
	10	0.9998	0.9999	0.9999	0.1555	0.3533	0.6452	0.0126	0.0061	0.0047
		0.0001	0.0000	0.0000	0.0466	0.0758	0.0721	0.0074	0.0003	0.0004
DTLZ5	4	0.7667	0.7671	0.7149	0.3674	0.5272	0.9688	0.0087	0.0094	0.0074
		0.0053	0.0041	0.1297	0.0657	0.0573	0.0654	0.0011	0.0009	0.0032
	6	0.8442	0.8405	0.8358	0.4253	0.7173	0.8715	0.0223	0.0148	0.0138
		0.0054	0.0068	0.0149	0.0694	0.0853	0.0826	0.0036	0.0014	0.0021
	8	0.8248	0.8054	0.8189	0.8396	1.0771	1.0687	0.0100	0.0057	0.0083
		0.0054	0.0174	0.0114	0.1396	0.0387	0.0599	0.0030	0.0014	0.0018
	10	0.8635	0.8321	0.8543	0.8545	1.0884	1.0731	0.0110	0.0066	0.0085
		0.0052	0.0233	0.0128	0.1432	0.0814	0.1087	0.0022	0.0013	0.0012
DTLZ6	4	0.9232	0.9217	0.8053	0.5981	0.6505	0.8747	0.0535	0.0500	0.0643
		0.0052	0.0090	0.1326	0.0721	0.0793	0.1894	0.0059	0.0055	0.0180
	6	0.9735	0.9742	0.9735	0.7450	0.8711	0.8668	0.0521	0.0401	0.0355
		0.0068	0.0046	0.0029	0.1167	0.0911	0.0866	0.0223	0.0175	0.0173
	8	0.9801	0.9796	0.9790	1.0136	1.1099	1.0374	0.0547	0.0163	0.0177
		0.0041	0.0032	0.0044	0.1128	0.0672	0.0935	0.0227	0.0078	0.0067
	10	0.9823	0.9808	0.9798	0.9752	1.1086	1.0948	0.0364	0.0179	0.0278
		0.0024	0.0012	0.0026	0.0960	0.0782	0.0871	0.0152	0.0051	0.0052
DTLZ7	4	0.1890	0.1815	0.1527	0.4350	0.5319	0.8111	0.0089	0.0084	0.0077
		0.0060	0.0084	0.0085	0.0708	0.0681	0.0697	0.0011	0.0011	0.0005
	6	0.1530	0.1522	0.1403	0.3081	0.6739	0.7942	0.0126	0.0126	0.0111
		0.0104	0.0148	0.0105	0.0542	0.1054	0.0791	0.0018	0.0011	0.0007
	8	0.5784	0.6050	0.6105	0.3834	0.6815	0.9145	0.0551	0.0155	0.0132
		0.0072	0.0211	0.0072	0.0586	0.0699	0.0424	0.0133	0.0011	0.0013
	10	0.0379	0.1368	0.1080	0.4258	0.6359	0.9397	0.0282	0.0138	0.0113
		0.0094	0.0106	0.0349	0.0881	0.0966	0.0790	0.0012	0.0011	0.0034

Table 2. Mean and standard deviation values of hypervolume results of the proposed method with ENS and A_ENS by incorporating them in existing NSGA-III

Problem	M	Hypervolume			Spread			GD		
		NSGA-III			NSGA-III			NSGA-III		
		ENS	AENS	ASNDS	ENS	AENS	ASNDS	ENS	AENS	ASNDS
DTLZ1	4	0.9122	0.8766	0.3786	0.0095	0.8833	1.1527	0.0021	0.0755	0.0018
		0.0005	0.0211	0.1307	0.0109	0.4733	0.1745	0.0000	0.1824	0.0027
	6	0.9800	0.6635	0.2470	0.1667	1.2243	1.1220	0.0144	0.1464	0.0005
		0.0019	0.2902	0.0839	0.2970	0.4333	0.0645	0.0622	0.3273	0.0005
	8	0.9300	0.3321	0.2604	0.4883	1.3017	1.0916	0.0649	0.0567	0.0007
		0.1693	0.2411	0.1004	0.5912	0.3373	0.0775	0.1793	0.1415	0.0020
	10	0.9564	0.3153	0.2610	0.5054	1.2594	1.0819	0.0702	0.0194	0.0005
		0.1018	0.1708	0.0956	0.6040	0.2927	0.0491	0.1654	0.0407	0.0016
DTLZ2	4	0.5704	0.5655	0.5197	0.1420	0.3593	0.8293	0.0055	0.0052	0.0031
		0.0008	0.0014	0.0122	0.0046	0.0489	0.0763	0.0001	0.0002	0.0003
	6	0.7459	0.7342	0.6753	0.1617	0.4433	0.9140	0.0048	0.0043	0.0041
		0.0008	0.0053	0.0475	0.0064	0.0637	0.0713	0.0001	0.0004	0.0006
	8	0.8349	0.8478	0.8237	0.3270	0.5182	0.9348	0.0052	0.0056	0.0048
		0.0678	0.0688	0.0741	0.2825	0.1728	0.1012	0.0004	0.0007	0.0007
	10	0.8497	0.9171	0.8975	0.4897	0.4563	0.9273	0.0055	0.0060	0.0051
		0.1124	0.0780	0.0683	0.3126	0.1055	0.0757	0.0017	0.0008	0.0006
DTLZ3	4	0.5679	0.4202	0.2451	0.1468	1.1576	1.2521	0.0050	0.3201	0.1606
		0.0047	0.1239	0.0786	0.0051	0.4953	0.2828	0.0000	0.6600	0.4797
	6	1.0000	0.9989	0.9980	0.9482	1.2726	1.0874	0.1058	0.4987	0.1582
		0.0000	0.0019	0.0027	0.1865	0.2236	0.1679	0.3266	0.5547	0.3192
	8	1.0000	1.0000	1.0000	1.2203	1.1755	1.0550	0.7172	0.3309	0.0874
		0.0000	0.0000	0.0000	0.2699	0.1102	0.0769	0.9587	0.3542	0.2259
	10	1.0000	1.0000	1.0000	1.2576	1.2218	1.0255	1.0615	0.2625	0.0070
		0.0000	0.0000	0.0000	0.2011	0.0853	0.0488	0.6874	0.2761	0.0151
DTLZ4	4	0.4253	0.3963	0.4366	0.6255	0.7845	0.9061	0.0041	0.0037	0.0028
		0.1467	0.1421	0.1447	0.4076	0.2591	0.1053	0.0013	0.0013	0.0004
	6	0.7126	0.6517	0.6519	0.5625	0.8261	0.9820	0.0050	0.0044	0.0039
		0.0882	0.1058	0.0975	0.4008	0.2324	0.1376	0.0006	0.0007	0.0010
	8	0.8147	0.8121	0.8258	0.6892	0.7671	0.8885	0.0046	0.0050	0.0044
		0.0730	0.0443	0.0601	0.4258	0.2237	0.2254	0.0007	0.0010	0.0009
	10	0.9124	0.9283	0.9237	0.6741	0.6415	0.8391	0.0049	0.0052	0.0044
		0.0494	0.0298	0.0370	0.4242	0.2825	0.2410	0.0016	0.0011	0.0014
DTLZ5	4	0.3909	0.3917	0.3885	1.0567	1.0689	1.0457	0.0012	0.0011	0.0014
		0.0044	0.0041	0.0082	0.0436	0.0494	0.0222	0.0003	0.0001	0.0008
	6	0.4559	0.4617	0.4561	1.0029	1.0744	1.0889	0.0017	0.0016	0.0025
		0.0026	0.0030	0.0057	0.0753	0.0646	0.0567	0.0005	0.0004	0.0011
	8	0.4755	0.4751	0.4706	0.9428	1.1019	1.1201	0.0021	0.0013	0.0046
		0.0065	0.0114	0.0153	0.0675	0.0661	0.0708	0.0003	0.0005	0.0022
	10	0.5641	0.5433	0.5350	0.7049	1.0869	1.0882	0.0037	0.0020	0.0042
		0.0135	0.0176	0.0317	0.1015	0.0468	0.0552	0.0009	0.0007	0.0016
DTLZ6	4	0.8687	0.8642	0.8491	0.9854	1.0097	1.1045	0.0064	0.0067	0.0073
		0.0009	0.0009	0.0046	0.0417	0.0689	0.1093	0.0006	0.0004	0.0017
	6	0.9702	0.9691	0.9542	0.8016	0.8895	1.0048	0.0220	0.0235	0.0193
		0.0070	0.0025	0.0264	0.0421	0.0555	0.0695	0.0077	0.0030	0.0057
	8	0.9741	0.9674	0.9480	0.7982	0.9795	1.0133	0.0172	0.0189	0.0169
		0.0065	0.0075	0.0233	0.0392	0.0426	0.0605	0.0108	0.0024	0.0057
	10	0.9750	0.9707	0.9546	0.7287	0.9848	1.0303	0.0146	0.0223	0.0174
		0.0030	0.0051	0.0129	0.0497	0.0475	0.0441	0.0030	0.0041	0.0055
DTLZ7	4	0.1872	0.1853	0.1585	0.6311	0.6700	0.9135	0.0097	0.0083	0.0068
		0.0041	0.0125	0.0137	0.0445	0.0438	0.0919	0.0046	0.0006	0.0008
	6	0.1411	0.1561	0.1468	0.6427	0.6504	0.8566	0.0154	0.0137	0.0122
		0.0079	0.0088	0.0134	0.0641	0.0507	0.0657	0.0040	0.0007	0.0009
	8	0.1057	0.1578	0.1472	0.5986	0.6727	0.9450	0.0144	0.0156	0.0130
		0.0216	0.0112	0.0168	0.0656	0.0655	0.1789	0.0014	0.0010	0.0021
	10	0.1243	0.1610	0.1457	0.5563	0.6177	0.8741	0.0166	0.0152	0.0119
		0.0242	0.0097	0.0193	0.0561	0.0738	0.1518	0.0020	0.0009	0.0023

From the experimental results presented in the Table 1 and 2, we can observe that the proposed method ASNDS when incorporated into the existing PDMOEAs, KnEA and NSGA-III, exhibits significant results in terms of the GD which emphasizes the improvement in the convergence of the algorithm. In other words, the proposed approach when incorporated into the PDMOEAs significantly improves the convergence of the PDMOEA. From the results presented in the Table 1, we can clearly notice that the proposed ASNDS, performing better for all the problems from DTLZ1-DTLZ7 in terms of GD indicator when incorporated in the KnEA algorithm. Whereas from the results presented in the Table 2, it is clear that for except the problem DTLZ5, the proposed method performs better corresponding to the GD indicator when incorporated in the NSGA-III algorithm.

In terms of the Spread, the proposed ASNDS is outperformed by the ENS approach but comparable to the AENS approach that indicates need for the improvement of the diversity performance when ASNDS is incorporated into PDMOEA. From the results presented in Table 1, we can observe that for the problems DTLZ2, DTLZ4 and DTLZ7 the proposed ASNDS is performing poorly in terms of spread indicator when incorporated in KnEA. When incorporated in NSGA-III, the proposed method performs worse in accordance to spread indicator for the problems, DTLZ2, DTLZ4 and DTLZ6-DTLZ7 as depicted in the Table 2. According to the Hypervolume results, we can observe that the proposed ASNDS performs comparable to the ENS and AENS approaches, which depicts that the proposed approach improves the convergence at the cost of the diversity.

4 Conclusion

In this paper, we propose an approximate Nondominated sorting approach based on sum of the normalized objectives (ASNDS) which requires only two objective comparisons to establish the dominance relation. We have compared the performance of the proposed approach with the existing Efficient NDS (ENS) and approximate ENS (AENS) by incorporating them into the existing PDMOEAs. The experimental results demonstrate that the proposed method improves the convergence of the algorithms and on the other hand affects the diversity. The proposed ASNDS reduces the computational complexity of the NDS approach as it only considers two objective comparisons irrespective of total number of objectives considered for optimization where the existing AENS uses three objective comparisons and other NDS approaches require all the objectives to establish dominance relation between the individuals. In future, we would like to include epsilon concept to the proposed approximate NDS approach to improve the diversity.

Acknowledgement. This study was supported by the BK21 Plus project funded by the Ministry of Education, Korea (21A20131600011).

References

1. Deb, K., Pratap, A., Agarwal, S., Meyarivan, T.: A fast and elitist multiobjective genetic algorithm: NSGA-II. IEEE Trans. Evol. Comput. **6**, 182–197 (2002)
2. Zhang, X., Tian, Y., Jin, Y.: A knee point-driven evolutionary algorithm for many-objective optimization. IEEE Trans. Evol. Comput. **19**, 761–776 (2015)
3. Zhou, A., Qu, B.-Y., Li, H., Zhao, S.-Z., Suganthan, P.N., Zhang, Q.: Multiobjective evolutionary algorithms: a survey of the state of the art. Swarm Evol. Comput. **1**, 32–49 (2011)
4. Li, M., Yang, S., Liu, X.: Shift-based density estimation for Pareto-based algorithms in many-objective optimization. IEEE Trans. Evol. Comput. **18**, 348–365 (2014)
5. Palakonda, V., Mallipeddi, R.: Pareto dominance-based algorithms with ranking methods for many-objective optimization. IEEE Access **5**, 11043–11053 (2017)
6. Shi, C., Chen, M., Shi, Z.: A fast nondominated sorting algorithm. In: 2005 International Conference on Neural Networks and Brain. ICNN&B 2005, pp. 1605–1610 (2005)
7. McClymont, K., Keedwell, E.: Deductive sort and climbing sort: new methods for non-dominated sorting. Evol. Comput. **20**, 1–26 (2012)
8. Wang, H., Yao, X.: Corner sort for Pareto-based many-objective optimization. IEEE Trans. Cybern. **44**, 92–102 (2014)
9. Zhang, X., Tian, Y., Cheng, R., Jin, Y.: An efficient approach to nondominated sorting for evolutionary multiobjective optimization. IEEE Trans. Evol. Comput. **19**, 201–213 (2015)
10. Palakonda, V., Pamulapati, T., Mallipeddi, R., Biswas, P.P., Veluvolu, K.C.: Nondominated sorting based on sum of objectives. In: 2017 IEEE Symposium Series on Computational Intelligence (SSCI), pp. 1–8 (2017)
11. Zhang, X., Tian, Y., Jin, Y.: Approximate non-dominated sorting for evolutionary many-objective optimization. Inf. Sci. **369**, 14–33 (2016)
12. Qu, B.-Y., Suganthan, P.N.: Multi-objective differential evolution based on the summation of normalized objectives and improved selection method. In: 2011 IEEE Symposium on Differential Evolution (SDE), pp. 1–8 (2011)
13. Deb, K., Jain, H.: An evolutionary many-objective optimization algorithm using reference-point-based nondominated sorting approach, part I: solving problems with box constraints. IEEE Trans. Evol. Comput. **18**, 577–601 (2014)
14. Huband, S., Hingston, P., Barone, L., While, L.: A review of multiobjective test problems and a scalable test problem toolkit. IEEE Trans. Evol. Comput. **10**, 477–506 (2006)
15. While, L., Hingston, P., Barone, L., Huband, S.: A faster algorithm for calculating hypervolume. IEEE Trans. Evol. Comput. **10**, 29–38 (2006)
16. Bringmann, K., Friedrich, T.: Approximation quality of the hypervolume indicator. Artif. Intell. **195**, 265–290 (2013)
17. Wang, Y.-N., Wu, L.-H., Yuan, X.-F.: Multi-objective self-adaptive differential evolution with elitist archive and crowding entropy-based diversity measure. Soft. Comput. **14**, 193 (2010)
18. Van Veldhuizen, D.A., Lamont, G.B.: Multiobjective evolutionary algorithm research: a history and analysis. Citeseer (1998)

Evolutionary Bi-objective Optimization and Knowledge Extraction for Electronic and Automotive Cooling

Shree Ram Pandey[1], Rituparna Datta[2(✉)], Aviv Segev[2],
and Bishakh Bhattacharya[1]

[1] Department of Mechanical Engineering, Indian Institute of Technology Kanpur,
Kanpur, Uttar Pradesh, India
{srpandey,bishakh}@iitk.ac.in
[2] Department of Computer Science, University of South Alabama,
Mobile, AL, USA
{rdatta,segev}@southalabama.edu

Abstract. The heat sink is one of the most widely used devices for thermal management of electronic devices and automotive systems. The present study approaches the design of the heat sink with the aim of enhancing their efficiency and keeping the material cost to a minimum. The above-mentioned purpose is achieved by posing the heat sink design problem as a bi-objective optimization problem where entropy generation rate and material cost are the two conflicting objective functions. The minimum entropy generation rate reduces irreversibilities inherent in the system, thus leading to improved performance, while the reduction in material cost ensures its economic feasibility. This bi-objective optimization problem is solved using Non-dominated Sorting Genetic Algorithm (NSGA-II) in the presence of geometric restrictions and functional requirements. Heat sinks with two different flow directions, namely flow-through air cooling system and impingement-flow air cooling system, are optimized to identify the best geometric and flow parameters. Subsequently a knowledge extraction exercise is carried out over non-dominated solutions obtained from the multi-objective optimization, to establish a relationship between the objective function and involved design parameters. The knowledge extracted has significant potential to simplify the calculations performed by thermal engineering experts in the selection of the heat sink for a specific application.

Keywords: Evolutionary computation and Bi-objective optimization ·
Knowledge extraction · Electronic and automotive cooling · Plate-Fin
Heat Sink

1 Introduction

Every electronic device and automotive system need to dissipate a certain amount of heat to maintain its temperature within its operational range. This

© Springer Nature Switzerland AG 2020
A. Zamuda et al. (Eds.): SEMCCO 2019/FANCCO 2019, CCIS 1092, pp. 79–92, 2020.
https://doi.org/10.1007/978-3-030-37838-7_8

activity of controlling the heat to be rejected is known as thermal management. One of the most significant ways to achieve thermal management is to reduce thermal resistance. This can be ensured by increasing the surface area of the interfacing surface between heat generating devices and the cooling medium. The simplest way to enhance cooling under cost, space, and weight constraints is to use a heat sink with the fin. Plate-Fin Heat Sinks (PFHS) are generally integrated into electronic devices that cool by blowing-out the heat. The PFHS is in principle a heat exchanger component that cools the device by dissipating heat to the surrounding cooling medium. The PFHS consists of two major parts - one part is a flat plate which is intended to make good thermal contact with the electronic device, and the other part is an array of comb-like protrusions to increase the surface area in contact with the cooling medium.

There have been several efforts to understand, apply, and improve the functioning of PFHSs. In the modern era, when every design has to pass through the philosophy of sustainable development, the design of the heat sink cannot be an exception. Bar-Cohen [2] observed that the sustainable development of the PFHS involves a subtle balance between a superior thermal design, minimum material consumption, and minimum pumping power. Bejan and Morega [4] introduced the concept of the minimization of the entropy generation. Culham and Muzychka [8] presented simultaneous optimization of PFHS design based on minimization of the entropy generation associated with heat transfer and fluid friction. Chen et al. [6] considered the minimization of entropy generation rate to be the objective function which is able to account for air resistance as well as the heat transfer resistance simultaneously. Mohsin et al. [12] applied Genetic Algorithm (GA) to minimize entropy generation rate due to heat transfer and pressure drop across pin fins. Culham et al. [7] highlighted the importance of the contribution made by all thermal resistance elements including contact resistance and spread resistance etc. between the heat source and the sink to the entropy generation. Ndao et al. [13] performed multi-objective [3] thermal design optimization and comparative study of various cooling technologies like continuous parallel micro-channel PFHSs, inline and staggered circular pin-fin PFHSs, offset strip fin PFHSs, and single and multiple submerged impinging jet(s). Mohsin et al. [12] used Genetic Algorithms (GAs) to minimize the entropy generation rate and demonstrated that geometric parameters, material properties, and flow conditions can be simultaneously optimized using GA. Sanaye and Hajabdollahi [14] carried out a thermo-economic optimization of the plate fin heat exchanger using genetic algorithms. A hybrid method was proposed by Ahmadi et al. [1] which is known as Genetic Algorithm Hybrid with Particle Swarm Optimization (GAHPSO) for design optimization of a plate-fin heat exchanger. The algorithm is able to handle both continuous and discrete variables. Another study of plate-fin heat exchangers was proposed by [11] using biogeography-based optimization (BBO). Wang and Li [17] proposed a method to address the problem of decrease in heat transfer performance and increase of pressure drop arising due to inappropriate surface selection and layer pattern. Ventola et al. [16] developed a novel thermal model of the PFHS, validated it experimentally, and demonstrated its superior accuracy.

The principle of the minimum entropy generation rate produces PFHS designs which are not only thermodynamically efficient but also have better geometric and topological features. This paper proposes a multi-objective optimization approach of PFHSs that are used as a cooling mechanism in electronic devices. The different variables of the optimization study are number of fins, height of the fin, spacing between the fins, and incoming air velocity. In addition to the restrictions on the lower and upper bounds of the design variables, there are also a few non-linear constraints from geometrical dependency, design specifications, and functional requirements. Two configurations such as PFHS with a flow-through air cooling system and PFHS with an impingement-flow air cooling system are considered in the present work. The PFHSs with impingement flow are used to obtain high local and area averaged heat transfer coefficients in the convective heat transfer process. Therefore, this configuration is used where heat flux density is significantly high, like cooling of turbine blades. The PFHS with flow through configuration is used when the constraint on the space availability is relatively relaxed. The conflicting objectives of multi-objective optimization are entropy generation rate and cost [9]. Minimum entropy generation rate will ensure better cooling. However, the solutions might not be economical. The minimum cost will ensure a design that works better from the economic point of view.

The structure and scope of the rest of the paper are organized as follows: Sect. 2 presents the details about the PFHS along with the heat sink design as a multi-objective problem defining design variables, constraints, and objective functions. The optimization results and the corresponding plots are discussed in Sect. 3. In Sect. 4, knowledge extraction methodologies are applied to the results obtained from the optimization study to establish a knowledge base for future reference. Finally, the concluding remarks and future development scope of the study are presented in Sect. 5.

2   Plate-Fin Heat Sink (PFHS)

There exists a large number of analysis tools for the determination of the thermal performance of PFHSs, provided design conditions are well defined. A model proposing a relationship between entropy generation and material cost with PFHS design parameters can be optimized in such a manner that relevant design parameters attain a value which combines to produce the best possible PFHS performance for a given set of constraints [15]. Two different configurations of a PFHS are considered in this paper. The first one is the flow-through air cooling system in Fig. 1 and the second one is the impingement-flow air cooling system in Fig. 2. The first flow configuration is used where a relatively large space is available as cooling fluid flows along the PFHS and not directly on the hot surface. The impingement-flow air cooling system (shown in Fig. 2) is suitable for the applications where large electronic component density exists and high heat flux needs to be dissipated. In this cooling arrangement, the goal is achieved by impingement of high velocity cooling fluid directly on the surface to be heated.

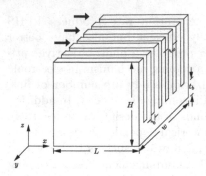

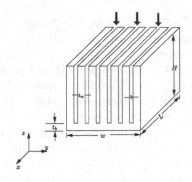

Fig. 1. PFHS with flow-through cooling.

Fig. 2. PFHS with impingement-flow cooling.

2.1 Multi-objective Optimization Problem Formulation

In the present work, an attempt has been made for simultaneous minimization of two conflicting objectives, the entropy generation rate (thermal performance) and the material cost (economy). This problem is adopted from Chen and Chen [5]. The multi-objective optimization design problem can be formulated as follows:

$$C_{mat} = (w \times L \times t_b + N \times H \times b \times L) \times \rho \times Cost.$$

$$\dot{S}_{gen} = R_{sink} \times \left(\frac{\dot{Q}}{T_{amb}}\right)^2 + \frac{F_d \times V_f}{T_{amb}}. \tag{1}$$

Where C_{mat} is the cost of material from which the PFHS is made, w width of the fin, L length of fins, t_b base length of fins, N number of fins, H height of the fin, b spacing between fins, ρ density of the cooling fluid, $\dot{S}_{gen}$ entropy generation rate, R_{sink} overall PFHS thermal resistance, $\dot{Q}$ represents heat generation rate, T_{amb} absolute surrounding temperature, F_d fluid friction in the form of drag force, and V_f uniform stream velocity. The overall PFHS resistance as defined in the case of flow-through air cooling systems and impingement-flow air cooling systems are given by:

$$R_{sink} = \begin{cases} \dfrac{1}{\left(\dfrac{N}{R_{fin}}\right) + (h_{eff} \times (N-1) \times b \times L)} \\ \quad + \dfrac{t_b}{k \times L \times w}, \text{ for flow-through air inlet,} \\ \dfrac{1}{h_{eff} \times A \times \eta_{fin}}, \text{ for impingement-flow air inlet.} \end{cases} \tag{2}$$

where R_{fin} is the thermal resistance of a single fin, h_{eff} is the effective heat transfer coefficient (the fins being assumed as straight fins with an adiabatic tip), k is the thermal conductivity, A is the total surface area of PFHS and other exposed surfaces, and η_{fin} represents the total heat dissipation efficiency. The objective functions are subjected to the following constraints:

$$g_1 : 0.001 - \left(\frac{w - t_w}{N - 1} - t_w \right) \leq 0,$$

$$g_2 : \left(\frac{w - t_w}{N - 1} - t_w \right) - 0.005 \leq 0,$$

$$g_3 : 0.001 - \left(\frac{H}{\left(\frac{w - t_w}{N - 1} \right) - t_w} \right) \leq 0,$$

$$g_4 : \left(\frac{H}{\left(\frac{w - t_w}{N - 1} \right) - t_w} \right) - 194.0 \leq 0,$$

$$g_5 : 0.0001 - \sqrt{\frac{\left(\frac{w - t_w}{N - 1} - t_w \right) V_{ch}}{\nu} \times \frac{\left(\frac{w - t_w}{N - 1} - t_w \right)}{L}} \leq 0.$$

$$(3)$$

The first two constraints, g_1 and g_2, put a limit on the fin gap, and according to these constraints the fin gap should lie in the range of 0.001 m to 0.005 m. The other two constraints deal with design specifications (g_3 and g_4) that arise due to limited space for installation. According to these constraints, the fin aspect ratio (ratio of height and thickness of the fin) should lie in the range of 0.01 and 194. The constraint g_5 is simply to avoid getting a zero Reynolds number. Beside these constraints, the design parameters can attain only those values which fall in the admissible limits. These admissible values of the design parameters are as follows:

$$
\begin{aligned}
2 &\leq N \leq 40, \\
0.014 &\leq H \leq 0.025, \\
2 \times 10^{-4} &\leq b \leq 2.5 \times 10^{-3}, \\
0.5 &\leq V_f \leq 2, \\
N &\times b \leq 0.05.
\end{aligned}
\qquad (4)
$$

3   Optimization Results

In the present work, an attempt has been made to simultaneously minimize of two conflicting objectives - the entropy generation rate (from the thermal performance perspective) and the material cost (from the economic perspective). It is observed from Eq. 2 that in case of flow through configuration, the thermal resistance of the PFHS is inversely proportional to number of fins. As increase in the number of fins also translates into enhanced exposed surface area, in case of impingement flow also, the inverse relationship between the thermal resistance and the number of fins remains valid. As a result, in both flow configurations increase in number of fins apparently leads to decrease in entropy generation rate ($\dot{S}_{gen}$). However, it is pertinent to note that the increase in number of fins also results in increased drag force being offered to the fluid flow. This increase in drag force has a consequential effect of increase in the entropy generation rate ($\dot{S}_{gen}$). The simultaneous interaction of both PFHS resistance and viscous dissipation must be taken into account in the PFHS optimization procedure in order to establish optimal operating conditions. All variables of interest, namely number of fins (N), height of the fin (H), spacing between fins (b), and incoming air velocity (v_f), have been constrained between their lower and upper bounds, hence providing a simultaneous optimization of all design variables. The multi-objective optimization problem of the PFHS is solved using Non-dominated Sorting Genetic Algorithm-II (NSGA-II [10]). Two different configurations of the PFHS (a) PFHS with a flow through air cooling system and (b) PFHS with an impingement flow air cooling system have been considered in the present work. The formulations of these two configurations are adapted from Chen and Chen [5] and the function evaluations are set similar to them. The following parameters are used for all the optimization tasks in the present work:

- Population size = 100,
- Generations = 100,
- Crossover probability (Simulated Binary Crossover) = 0.9,
- SBX index = 10,
- Mutation (Polynomial mutation) probability = 1/number of variables,
- Mutation index = 100.

The Pareto-optimal solutions obtained from NSGA-II are shown in Fig. 3. Figure 3 consists of the comparative performance of our method with Chen and Chen [5] for a PFHS with a flow through air cooling system. The non-dominated solutions in the figure clearly show that the performance of the present method is better than results by Chen and Chen [5].

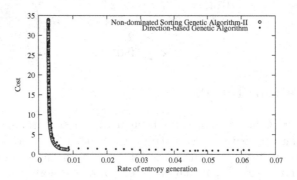

Fig. 3. Non-dominated solutions between rate of entropy generation and cost for flow-through

The entropy generation rate varies between 0.002898 W/k and 0.008558 W/k. The lowest cost is 1.132713NTD whereas the highest cost is 33.920260NTD.

Table 1. Non-dominated solutions along with variable values for PFHS with flow through

Rate of entropy generation (W/k)	Cost (NTD)	Number of fins	Height of the fin (mm)	Spacing between fins (mm)	Incoming air velocity (m/s)
0.002898	33.920260	18	0.124923	0.001144	1.186749
0.002934	21.023040	22	0.089170	0.001130	1.507174
0.003002	14.888590	25	0.072592	0.001143	1.757355
0.003149	9.480572	29	0.055671	0.001135	2.0
0.003552	5.350479	36	0.043285	0.001077	2.0
0.004296	2.956553	40	0.032595	0.001084	2.0
0.005281	1.876195	40	0.027582	0.001169	2.0
0.006376	1.420160	40	0.025001	0.001214	1.999995
0.007512	1.227456	40	0.025	0.001238	2.0
0.008558	1.132713	40	0.025	0.001250	1.999499

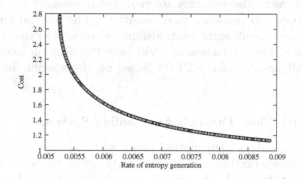

Fig. 4. Non-dominated solutions for rate of entropy generation and cost for impingement-flow

Table 1 shows two extreme values and some intermediate values of the objective function along with corresponding design variables. It can also be observed from Table 1 the number of fins is directly proportional to the rate of entropy generation. The number of fins attains its highest value (40) when the entropy generation rate ranges from 0.004296 W/k to 0.008558 W/k. The incoming velocity achieves its highest value (2 m/s) when entropy generation rate reaches 0.003149 W/k. We can observe that with further increase, the number of fins and incoming air velocity entropy generation will increase.

The resulting non-dominated solutions between rate of entropy generation and cost for the PFHS with an impingement flow air cooling system are shown in Fig. 4. The entropy generation rate ranges from 0.005247 W/k to 0.008879 W/k and the cost varies between 1.132710NTD and 2.771404NTD. The extreme and intermediate values of objective functions for this configuration along with design variables are shown in Table 2. The table clearly shows that number of fins always takes its highest bound (40) and height of the fin is always 0.025 mm. The incoming air velocity is fixed at its upper bound i.e. 2 m/s. Table 2 also shows that three variables out of four are fixed for all non-dominated solutions. The only variable that causes different non-dominated solutions is spacing between the fins.

In the present work, in addition to solving the above mentioned two configurations, knowledge extraction has been carried out from the obtained solutions of multi-objective optimization. The motivation of the knowledge extraction is to establish a relationship between input design variables and output non-dominated solutions of the multi-objective optimization problem. This knowledge will help the user to select the fin for specific application.

4   Knowledge Extraction

The non-dominated solutions of multi-objective optimization were shown in the previous section. Next, the solutions are analyzed thoroughly to extract knowledge from the obtained non-dominated solutions. The motivation is to establish the existence of meaningful relationships between objective functions and decision variables. These relationships will help the decision maker select the appropriate configuration of the PFHS based on the specific and customized needs.

4.1   PFHS with Flow Through Air Cooling System

All the decision variables were plotted along with the first objective function (rate of entropy generation) to visualize the relationships between objective functions and design variables. As the two objective functions considered in this case are conflicting, it would be sufficient to establish the relationship of the design variables with any one of the objective functions. The dependence on the other objective function with design variables can be predicted by exploiting the fact that both the objective functions are conflicting. However, it can be argued that analytical relationships between the objective functions and the design variables are obtained for both the objective functions separately so that the exact dependence on the design variables can be understood.

The change in the rate of entropy generation ($\dot{S}_{gen}$) with the variation in the number of fins (N) is shown in Fig. 5. Existence of two distinct zones is visible in Fig. 5. The first zone shows that as $\dot{S}_{gen}$ decreases, there is a corresponding decrease in number of fins. This plot also gives the exact relationship between $\dot{S}_{gen}$ and N in Zone 1 as shown in Eq. 5:

$$N = -1963.96 + (1.69185e^6 \dot{S}_{gen}) - (4.79851e^8 \dot{S}_{gen}^2) + (4.56218e^{10} \dot{S}_{gen}^3). \quad (5)$$

In Zone 2, the number of fin attains its maximum allowable value of 40 corresponding to a critical value of entropy generation rate, the value being

Table 2. Non-dominated solutions along with variables for PFHS with impingement flow.

Rate of entropy generation (W/k)	Cost (NTD)	Number of fins	Height of the fin (mm)	Spacing between fins (mm)	Incoming air velocity (m/s)
0.005247	2.771404	40	0.025	0.001042	2.0
0.005380	2.173199	40	0.025	0.001118	2.0
0.005551	1.935973	40	0.025	0.001148	2.0
0.005828	1.717386	40	0.025	0.001176	2.0
0.006183	1.548827	40	0.025	0.001197	2.0
0.006628	1.414482	40	0.025	0.001214	2.0
0.007165	1.308253	40	0.025	0.001228	2.0
0.007726	1.232049	40	0.025	0.001237	2.0
0.008302	1.175486	40	0.025	0.001245	2.0
0.008879	1.132710	40	0.025	0.001250	2.0

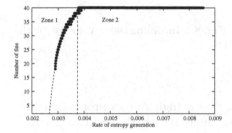

Fig. 5. Variation of number of fins with $(\dot{S}_{gen})$

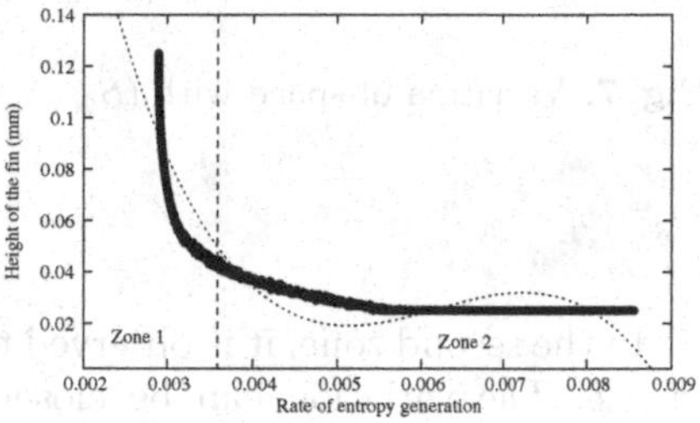

Fig. 6. Variation of height of fin with $(\dot{S}_{gen})$

0.0039 w/k. Once this critical value is achieved, there is no change in the number of fins in the design with any further increment in the entropy generation rate. It should be noted that if any design calculation gives N as a non-integer value, it should be approximated with the nearest integer value. In the second zone, the number of fins is always fixed at its upper bound.

Figure 6 shows variation in height of fin (H) with entropy generation rate $(\dot{S}_{gen})$ as obtained from the Pareto-optimal solutions of optimization results. The two zones do not have very clear distinction in Fig. 6. However, two different zones have been identified to have uniformity in the discussion. In the first zone, the variation in H with entropy generation rate, $\dot{S}_{gen}$, has a steep slope. However, the variation in H with respect to $\dot{S}_{gen}$, is very minimal in the second zone compared to the first zone. The relationship plot can be approximated by the cubic polynomial (Eq. 6):

$$H = 0.618 - 305.676\dot{S}_{gen} + 50881.1\dot{S}_{gen}^2 - 2.741e + 06\dot{S}_{gen}^3. \qquad (6)$$

Figure 7 shows the relationship between variation of entropy generation rate $(\dot{S}_{gen})$ and spacing between the fins (b). The observation can be divided into two different zones. The first zone ranges from 0.002 W/k to 0.0038 W/k whereas the second zone lies from 0.0039 W/k to 0.86 W/k. An inverse proportionality exists in the first zone. The variation in the first zone can be approximated using a linear equation of the form as given in Eq. 7 (Zone 1).

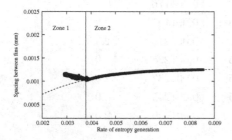

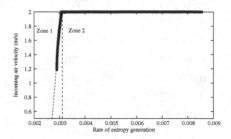

Fig. 7. Variation of space with $(\dot{S}_{gen})$

Fig. 8. Incoming air velocity with $(\dot{S}_{gen})$

In the second zone, it is observed that $\dot{S}_{gen}$ bears higher order proportionality with b. The variation can be closely approximated with the help of a cubic polynomial of the following form (Eq. 7 (Zone 2)):

$$b = \begin{cases} 0.0015 - 0.1106\dot{S}_{gen}, & \text{Zone 1,} \\[2ex] 7.32236e - 05 + 0.41382\dot{S}_{gen} \\ -50.2806\dot{S}_{gen}^2 + 2105.77\dot{S}_{gen}^3, & \text{Zone 2.} \end{cases} \qquad (7)$$

The junction of Zone 1 and Zone 2 shows zeroth order continuity where values are continuous but the derivatives are discontinuous. Figure 8 shows the variation of entropy generation rate $(\dot{S}_{gen})$ with incoming air velocity (v_f) as obtained from the result of post optimal analysis. The variation can be classified into two distinct zones. However, these two zones are dissimilar to the other three zones. In the first zone the entropy generation rate, $\dot{S}_{gen}$, varies linearly with incoming air velocity, v_f. This linear variation has a very high slope indicating that for a small change in entropy generation rate there is significant change in incoming air velocity. The linear variation ceases to exist at the critical value of entropy generation rate, which is 0.0033 W/k. The incoming air velocity attains its allowable maximum limit of $2 m/s$ at the critical rate of entropy generation of 0.0033 W/k and after that it remains unchanged with further variation in entropy

generation rate in Zone 2. The analytical relationship between entropy generation rate $(\dot{S}_{gen})$ and incoming air velocity (v_f) in Zone 1 can be approximated
as follows (Eq. 8):

$$v_f = -11.076 + 4269.79\dot{S}_{gen} \tag{8}$$

4.2 PFHS with Impingement Flow Air Cooling System

In an interesting observation, the rate of entropy generation $(\dot{S}_{gen})$ varies only
with fin spacing parameter (b) while being invariant with the other three design
variables (Figs. 9(a), (b), and 11). To analyze how one optimal solution differs
from the other optimal solutions, all four design parameters have been plotted
against rate of entropy generation, $\dot{S}_{gen}$ (Figs. 9 and 10).

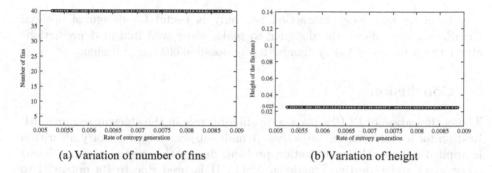

(a) Variation of number of fins   (b) Variation of height

Fig. 9. Variation of number of fins and height of fin with rate of entropy generation
$(\dot{S}_{gen})$

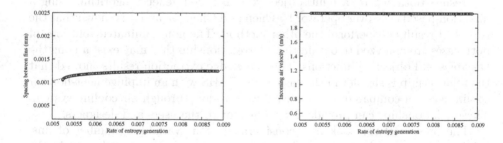

Fig. 10. Variation of space with $(\dot{S}_{gen})$ **Fig. 11.** Incoming air velocity with
$(\dot{S}_{gen})$

For a lower entropy generation rate, the fin spacing parameter should be
small and with any increase in entropy generation rate, the fin spacing parameter increases monotonically with $(\dot{S}_{gen})$. This variation can be explained from

the practical observation that for a fixed plate dimension a small fin spacing parameter would result in a larger number of fins. This increase in the number of fins would eventually result in a lower entropy generation rate. Therefore, if better fin performance is desired, it is advisable to design a PFHS with a lower value of fin spacing parameter and to keep all the other design variables at their prescribed constant values as discussed below. Hence, it can be inferred from this knowledge extraction methodology that if an optimal PFHS is to be designed with four design parameters $(N, H, b,$ and $v_f)$, the three parameters namely $N, H,$ and v_f must be fixed whereas b can be adjusted to obtain the desired trade off among various chosen objectives. The relationship between the rate of entropy generation ($\dot{S}_{gen}$) and b from Fig. 10 is shown below (Eq. 9):

$$b = -0.0002 + 0.367\dot{S}_{gen} - 23.4294\dot{S}_{gen}^2 \tag{9}$$

Therefore, knowledge extraction not only is useful for design of optimal PFHS, but also allows the designer to make some well informed predictions about the behavior of the system with any possible change in design.

5   Conclusion

The optimization of PFHSs plays a meaningful role in the efficient resource utilization for a given cooling objective. A multi-objective evolutionary algorithm is applied to solve the optimization problem due to the existence of non-linear constraints and objective functions. NSGA-II is used due to its potential to deal with non-linear constraints and objective functions in multi-objective optimization problems. It is evident from non-dominated solutions that NSGA-II has successfully generated well-spread non-dominated solutions. The non-dominated solutions of a PFHS with a flow-through air cooling system are compared with the results obtained by a multi-objective real-coded genetic algorithm using a direction-based crossover operator by Chen and Chen [5], and it is shown that the NSGA-II results outperform the other method. The non-dominated solutions of both cases are analyzed to obtain the interrelationship that may exist among the variables and objective functions. The knowledge extraction results showed that the relationship is simpler in the case of the PFHS with an impingement-flow air cooling system compared to the PFHS with a flow-through air cooling system. These relationships can provide a deep insight to the users and designers.

The present study took into consideration four variables (number of fins, height of fin, space between number of fins, and incoming air velocity) in the optimization study. Another direction of research could be to increase the number of constraints and objective functions and solve for both configurations as many objective optimization problems. A generalized formulation for the above two configurations can be designed which can assist designers in the development of a PFHS used for cooling electronic devices based on their specific needs in terms of cooling rate, space availability, and material cost. The PFHSs can also be integrated with smart materials to introduce adaptability in their geometry and performance. This would enable fins to vary their geometry and heat flux rate in response to change in the value of the thermal parameters of the surrounding cooling medium. The present study can further be used to find the right combination of conventional and smart materials to yield the optimal value of thermal performance, material cost, and operational cost.

References

1. Ahmadi, P., Hajabdollahi, H., Dincer, I.: Cost and entropy generation minimization of a cross-flow plate fin heat exchanger using multi-objective genetic algorithm. J. Heat Transf. **133**(2), 021801 (2011)
2. Bar-Cohen, A.: Thermal packaging for the 21st century: challenges and options. In: Fifth Therminic-International Workshop Thermal Investigations of ICs and Systems, pp. 3–6 (1999)
3. Bechikh, S., Datta, R., Gupta, A. (eds.): Recent Advances in Evolutionary Multi-objective Optimization. ALO, vol. 20. Springer, Cham (2017). https://doi.org/10.1007/978-3-319-42978-6
4. Bejan, A., Morega, A.M.: Optimal arrays of pin fins and plate fins in laminar forced convection. J. Heat Transf. **115**(1), 75–81 (1993)
5. Chen, C.T., Chen, H.I.: Multi-objective optimization design of plate-fin heat sinks using a direction-based genetic algorithm. J. Taiwan Inst. Chem. Eng. **44**(2), 257–265 (2013)
6. Chen, C.T., Wu, C.K., Hwang, C.: Optimal design and control of cpu heat sink processes. IEEE Trans. Compon. Packag. Technol **31**(1), 184–195 (2008)
7. Culham, J.R., Khan, W.A., Yovanovich, M.M., Muzychka, Y.S.: The influence of material properties and spreading resistance in the thermal design of plate fin heat sinks. J. Electron. Packag. **129**(1), 76–81 (2007)
8. Culham, J.R., Muzychka, Y.S.: Optimization of plate fin heat sinks using entropy generation minimization. IEEE Trans. Compon. Packag. Technol. **24**(2), 159–165 (2001)
9. Datta, R., Pandey, S.R., Segev, A.: Multi-objective optimization and knowledge extraction of plate-fin heat sink using evolutionary algorithms. In: Global Optimization Conference (GOC-2017). The World Congress on Global Optimization (WCGO) (2017)
10. Deb, K., Agrawal, S., Pratap, A., Meyarivan, T.: A fast and elitist multi-objective genetic algorithm: NSGA-II. IEEE Trans. Evol. Comput. **6**(2), 182–197 (2002)
11. Hadidi, A.: A robust approach for optimal design of plate fin heat exchangers using biogeography based optimization (BBO) algorithm. Appl. Energy **150**, 196–210 (2015)
12. Mohsin, S., Maqbool, A., Khan, W.A.: Optimization of cylindrical pin-fin heat sinks using genetic algorithms. IEEE Trans. Compon. Packag. Technol. **32**(1), 44–52 (2009)
13. Ndao, S., Peles, Y., Jensen, M.K.: Multi-objective thermal design optimization and comparative analysis of electronics cooling technologies. Int. J. Heat Mass Transf. **52**(19), 4317–4326 (2009)
14. Sanaye, S., Hajabdollahi, H.: Thermal-economic multi-objective optimization of plate fin heat exchanger using genetic algorithm. Appl. Energy **87**(6), 1893–1902 (2010)

15. Sharma, A.K., Datta, R., Elarbi, M., Bhattacharya, B., Bechikh, S.: Practical applications in constrained evolutionary multi-objective optimization. In: Bechikh, S., Datta, R., Gupta, A. (eds.) Recent Advances in Evolutionary Multi-objective Optimization. ALO, vol. 20, pp. 159–179. Springer, Cham (2017). https://doi.org/10.1007/978-3-319-42978-6_6
16. Ventola, L., et al.: Unshrouded plate fin heat sinks for electronics cooling: validation of a comprehensive thermal model and cost optimization in semi-active configuration. Energies **9**, 608 (2016)
17. Wang, Z., Li, Y.: A combined method for surface selection and layer pattern optimization of a multistream plate-fin heat exchanger. Appl. Energy **165**, 815–827 (2016)

Classification of Stock Market Trends with Confidence-Based Selective Predictions

Wen Xin Cheng[✉][ID], P. N. Suganthan[ID], Xueheng Qiu, and Rakesh Katuwal

School of Electrical and Electronic Engineering, Nanyang Technological University, 50 Nanyang Avenue, Singapore 639798, Singapore
{wenxin001,epnsugan,qiux0004,rakeshku001}@ntu.edu.sg

Abstract. Predicting the trend of stock price movement accurately allows investors to maximize their profits from investments. However, due to the complexity of the stock data, classifiers often make errors, which cause the investors to lose money from failed investments. This study attempts to reduce such risks by focusing on easy-to-classify cases that have the highest chances of success. Therefore, we propose a method which selects only the predictions that have the highest confidence. In an experiment on 50 stocks, each learning model is trained on each stock data and evaluated based on the classification accuracy over a moving time window. The models which have the highest confidence are selected to predict the trend for that stock the next day. The experiment results shows the classification accuracy has improved significantly when the top 10% of predictions were used.

Keywords: Stock price forecasting · Time series classification · Ensemble methods

1 Introduction

Over the years, the popularity of machine learning for stock price forecasting has gathered more interest from economic experts, traders and companies in the financial market. By analyzing the historic stock data, experts can recognize patterns and trends which helps investors to make good investment strategies. However, forecasting stock prices movement is an extremely challenging task even if the task is simplified into 2 classes ('up' or 'down'). Stock market prices can be affected by many factors, including economy condition, government policies and the interest of investors [1,11]. As a result, stock data are often complex in nature and contain highly non-linear and non-stationary patterns, which makes the time series difficult to classify without using other informative sources. In addition, the large amount of noise present in such time series poses a big challenge for forecasting the trend of its movement.

A number of stock forecasting methods have been proposed with varying levels of success. Statistical-based time series methods such as linear regression [23],

© Springer Nature Switzerland AG 2020
A. Zamuda et al. (Eds.): SEMCCO 2019/FANCCO 2019, CCIS 1092, pp. 93–104, 2020.
https://doi.org/10.1007/978-3-030-37838-7_9

Holt-Winters exponential smoothing [14] and Autoregressive Integrated Moving Average (ARIMA) [17] uses classical statistical theories and mathematical formulas to predict the trend of the stock price.

Machine learning methods learns from the time series signals and make data-driven predictions. Such models includes Support Vector Machines (SVM) [5], Support Vector Regression (SVR) [8], Artificial Neural Networks (ANN) [6], and Random Forests [3].

Recently, works on Deep Neural Networks has published to explore the effectiveness of such algorithms on stock price forecasting. Such algorithms include Long Short Term Memory (LSTM) [13], Convolutional Neural Network (CNN) [20] and Deep Belief Network (DBN) [12]. In [7], the authors uses CNN and neural tensor network to model the relationship between influencing events and the stock price.

Random Vector Functional Link (RVFL) [21, 22] network is a type of neural networks which randomizes the weights instead of optimizing them. Since RVFL optimizes the output weights using close-form solution in place of the Backward Propagation, a lot of training time is saved, allowing the model to be trained at a much shorter time than traditional neural network. A similar method, single hidden layer neural network with random weights (RWSLFN) [32], is different from RVFL by removing the direct links. However, [30, 35] shows that including the direct link can significantly improve the performance of the classifier.

Ensemble learning methods combines several techniques together to improve the classification capabilities. Such methods has two different strategies to combine the results of each model: sequential and parallel [31]. In a sequential ensemble model, the output of several models is connected as inputs the next model [2, 26]. On the other hand, a parallel ensemble model first breaks the signal into several sub-signals. A model is then built on each sub-signal to perform classifications. Such methods include Empirical Mode Decomposition (EMD) [16] and Wavelet Decomposition [10, 15]. Two different ensemble models have been proposed to perform electrical load forecasting, one combining EMD with DBN [27] and another combining Discrete Wavelet Transform (DWT), EMD and RVFL [29]. In [28], authors uses a DWT-EMD-RVFL-SVR model to forecast stock prices of various power related companies.

While such state-of-the-art classifiers can be accurate in forecasting the trend of stock price movements, classifiers often misclassify the trend of the stock price movements due to the complexity of the historical stock price data. This would cause stock investors to lose money when they invest on stocks that are wrongly classified as 'increase'. One strategy to reduce such risks is to make forecasts on a wide range of stocks and then keep only the forecasts (say 10% of the total number of forecasts) that has the highest chance of success.

This work presents an new approach which reduces the risks of making wrong classifications by focusing on easy-to-classify stocks. For each stock, a learning model will be trained and evaluated based on the classification accuracy over a moving time window. The models which has the highest accuracy are then selected to predict the trend for that stock the next day. The efficiency of the proposed approach and the effect of selecting only the best classifiers are explored with the experiment on 50 United States stocks.

The rest of the paper is organized as follows: Sect. 2 describes the theoretical background of various methods. Our proposed method is described in Sect. 3. Section 4 presents our experiment setup and the results. Finally, Sect. 5 concludes the paper.

2 Theoretical Background

2.1 Wavelet Transform

Wavelet Transform [9] is a decomposition method which works in a similar way as Fourier Transform. The main difference between the two transformations is the representations. Wavelet Transform uses a library of wavelet functions to represent the original time series while the traditional Fourier Transform expresses the time series as a set of sine and cosine waves [19].

A variant of Wavelet Transform, Maximal Overlap Discrete Wavelet Transformation (MODWT) [24], decomposes the time series signal using high-pass and low-pass filters for every level. This variant has some advantages over the original Wavelet Transform. Firstly, MODWT is highly redundant and non-orthogonal, which enables better comparison between the time series signal and their decomposition [25]. In addition, MODWT is well defined in all sample sizes unlike Wavelet Transform, which is only defined when the sample size is a multiple of j levels [19,25].

2.2 Empirical Mode Decomposition

Empirical Mode Decomposition (EMD) [16], also known as Hilbert-Huang transform, was developed based on the assumption that every signal comprises of several simple intrinsic oscillation modes known as Intrinsic Mode Functions (IMF), Each IMF has only one zero-crossing between two local extrema. The residue signal is obtained by subtracting all the IMFs from the main signal.

Since stock price forecasting is a complex problem where the time series signal comprises of many individual components. Therefore, EMD can be applied to improve the classification performance. The original time series can be easily reproduced by summing all the IMFs and the residue.

However, "Mode Mixing Problem" is common for EMD decomposition and researchers are working on ways to fix this issue. Such works includes combining EMD and DWT sequentially such as the ensemble models in [28,29].

2.3 Random Vector Functional Link

Random Vector Functional Link (RVFL) is one variant of artificial neural networks which differs from traditional neural network in 3 ways: the input layer is connected directly to the output layer, the weights between the input layer and its hidden layers are randomized and a closed form least squares method is used in place of a back propagation method [21,22].

Each neuron in the hidden layer takes a weighted sum of the inputs and applies an activation function to produce an output, which is then used in the output layer. The weights of its hidden layer W_h are fixed at a random value between K and $-K$, where K is a tune-able parameter. With W_h and the activation function fixed, the weights of the output layer can be easily estimated using the Least Squares method.

2.4   Stock Market Indicators

Stock Market Indicators (SMI) are indicators which aids in the prediction of stock market movements. Such indicators are first used in [18] as features to classify Istanbul Stock Exchange and later used in [28] as part of the feature set of the Support Vector Regression.

From the works from [28], the authors mentioned 4 important indicators for stock price forecasting: Moving Average Convergence Divergence (MACD), Relative Strength Index (RSI), Stochastic Oscillator and Commodity Channel Index (CCI). The mathematical formulas of 10 stock market indicators used in this work are given in [18].

2.5   DWT-EMD-RVFL-SVR Model

The DWT-EMD-RVFL-SVR Model is an ensemble technique which combines several techniques to predict future values [28]. This method employ a divide-and-conquer concept where the algorithm first divides the complex problem into several smaller problems and then build prediction models to solve smaller problems.

Firstly, the closing stock price are decomposed by using both MODWT and EMD, where MODWT may aid in preventing the "Mode Mixing Problem" commonly present for EMD-decomposed signals [4,34]. Next, each decomposed signal are modeled using RVFL to perform predictions on the decomposed signal. The outputs of each RVFL model are then combined with 10 SMIs and the current closing price to train the Support Vector Regression (SVR) to predict the closing price for the next day.

3   Proposed Method

Stock price forecasting is often a challenging task with many factors affecting the direction of stock price movements. While data extracted from historical stock data can be very useful, historical stock data are often complex or shows no relationships towards future trends. In the worst case scenario, no useful information can be extracted and the classifier can only achieve approximately 50% accuracy. In such cases, predictions can be as good as random guessing.

To reduce the risk of misclassification, we propose a method which focuses on predicting the easy-to-classify cases. The main idea of this method is to train prediction models for a large number of stocks (50 stocks in this work), and consider only a few stocks (say 10% of the 50 stocks) with the highest confidence of making the correct prediction. The remaining 90% of the stocks are ignored. In this way, investors can choose to forecast only on stocks that has the highest chance of success.

For this work, we look at 50 US stocks. The historical data of each stock are treated as separate datasets. In each stock, a classifier will be trained and evaluated on the historical data of this stock. Since we looked at 50 stocks, this will generate 50 classifiers, with each classifier trained on each stock.

As for the method used in forecasting the direction of stock price movements, it is important to use an accurate forecasting model for the proposal to work. For this, the ensemble method proposed in [28] was used in this study as the ensemble method performed better than other state-of-the-art methods.

Since the forecasting model predicts the closing price for the next day x'_{t+1}, the predictions made by each model will then be converted to class labels y'_t based on the equation below:

$$y'_t = \begin{cases} +1, \, if \, x'_{t+1} > x_t \\ -1, \, otherwise \end{cases} \tag{1}$$

where x'_{t+1} is the predicted closing price for the next day and x_t is the current closing price.

Next, for each day, each model will be evaluated using samples taken from the past D days, where D is the length of the moving time window (time window is set to $\{50, 100, 150, 200, 250, 300\}$ days in this work). The trained models will be ranked based on their classification accuracy in the descending order. This means the model with the highest accuracy will be ranked 1 and the model with lowest accuracy will be ranked 50. Once the ranking is obtained, we pick the best classifiers (we use the top $\{25, 20, 10, 5, 4, 3, 2, 1\}$ classifiers in this work) to predict the direction of stock price movement for the next day. Table 1 shows the algorithm of our proposed method.

Algorithm 1. Algorithm for our Proposed Method

Input: Trained Model, Input data X, Labels Y and Predicted Train Labels Y'
Output: Predicted Test labels $y'_{t+1}, ...$
1: **for** Each Day t **do**
2:     **for** Each Stock **do**
3:         Predict the next day's closing price x'_{t+1}.
4:         If $x'_{t+1} > x_t$, set $Y'_{t+1} = 1$. Otherwise, set $Y'_{t+1} = 0$.
5:         Evaluate the accuracy of the classifier using samples from the past D days.
6:     **end for**
7:     Rank the models based their accuracy in the descending order.
8: **end for**

---

4 Experiment

4.1 Datasets

The closing prices from 50 United States stocks are used to analyze the effectiveness of the algorithms[1].

For each stock, daily stock prices from 08/12/2016 to 07/12/2018 are used. Each stock contains the closing price for the same set of days. The training and testing splits is performed as follows: The first 70% of the data is used for training while remaining 30% of data is used for testing.

In this experiment, each sample are labeled as one of the 2 classes based on the closing price of the stock. Each sample is labeled as '+1' if the closing price increases the next day and '−1' otherwise (See Subsect. 4.2). The class distribution of 50 datasets are recorded in Table 1.

4.2 Experimental Setup

In this experiment, we will be following the experiment setup based on [28]. Classifiers are tasked to predict whether the closing price increase the next day. We set the input vector $X_t = [x_{t-25}, x_{t-24}, ..., x_t]$ as the closing stock prices for the past 26 days. The corresponding class label Y_t is set to '+1' if the closing price for the next day is higher than the closing price for day t and '−1' otherwise. In other words, the corresponding class label is set based on the equation below.

$$y_t = \begin{cases} +1, \; if \; x_{t+1} > x_t \\ -1, \; otherwise \end{cases} \tag{2}$$

where x_{t+1} is the closing price for the next day and x_t is the closing price of day t. All data points in each dataset are linearly scaled down to range $[0, 1]$.

The ensemble learning method was implemented based on the works in [28]. Experiments are conducted using MATLAB. To test the efficiency our proposed method, we select the predictions with top $\{50, 25, 15, 10, 5, 4, 3, 2, 1\}$ confidence. Selecting the top 50 models indicates that no selection has been done (which means all stocks are considered). To test the effect of selecting different time windows, we repeat the experiment with the time window set to the past $\{50, 100, 150, 200, 250, 300\}$ days.

In addition to the classification accuracy, F1 Score are used in this paper. This performance matrix is useful in this study as it gives emphasis in getting true positives. Suppose investors were to make investments whenever they forecasts an increasing trend. True positives (TP) indicate a good investment, false positives (FP) indicate a poor investment and false negatives (FN) indicates a wasted opportunity. The definition is stated in the equation below [33]:

$$F1score = \frac{2 \times TP}{2 \times TP + FN + FP} \tag{3}$$

---

[1] The stock prices can be downloaded from Yahoo Finance at http://www.finance.yahoo.com/.

Table 1. Class distribution of 50 datasets

Stock	Train samples		Test samples		Stock	Train samples		Test samples	
	Positive	Negative	Positive	Negative		Positive	Negative	Positive	Negative
AAPL	193	140	111	33	MA	162	171	104	40
ABB	160	173	63	81	MSD	149	184	59	85
ADBE	173	160	110	34	MSFT	191	142	120	24
AT	99	234	44	100	MSI	155	178	69	75
AMD	156	177	66	78	NATI	155	178	57	87
AMZN	142	191	61	83	NDAQ	152	181	70	74
BA	184	149	122	22	NFLX	209	124	125	19
BB	171	162	73	71	NOK	160	173	65	79
CAJ	163	170	66	78	NVDA	245	88	142	2
CVX	149	184	77	67	ORCL	146	187	66	78
DAL	159	174	64	80	PEP	152	181	64	80
DOX	150	183	63	81	PRU	164	169	71	73
DVMT	138	195	56	88	PTR	160	173	87	57
F	156	177	76	68	QCOM	155	178	69	75
FB	155	178	68	76	RDS-B	148	185	69	75
GE	161	172	85	59	S	179	154	67	77
GOOGL	158	175	62	82	SGTZY	120	213	60	84
HMC	151	182	74	70	SNE	145	188	71	73
HP	161	172	82	62	STX	143	190	60	84
HPQ	269	64	142	2	T	160	173	69	75
IBM	163	170	75	69	TSLA	174	159	61	83
INTC	156	177	66	78	V	210	123	138	6
K	161	172	73	71	VZ	153	180	76	68
KINS	160	173	64	80	WDC	164	169	68	76
LOGI	143	190	59	85	XOM	160	173	74	70

Positive Class indicates closing price increases the next day
Negative Class indicates closing price does not increase the next day

4.3   Results and Discussion

Before analyzing the efficiency of our proposed method, we first check the classification performance of the classifier on individual stocks to determine if the classifier has any potential of delivering good classification performance. The classification performance for on 50 stocks is presented in Table 2.

Table 2. Classification performance on 50 stocks

Stock	Accuracy (%)		F1 score		Stock	Accuracy (%)		F1 score	
	Training	Testing	Training	Testing		Training	Testing	Training	Testing
AAPL	61.0	71.5	0.688	0.827	MA	61.6	70.8	0.602	0.828
ABB	62.2	52.1	0.616	0.601	MSD	61.6	45.8	0.584	0.512
ADBE	62.5	76.4	0.625	0.863	MSFT	64.0	81.9	0.707	0.901
AT	60.4	54.2	0.557	0.353	MSI	56.8	48.6	0.463	0.630
AMD	64.0	55.6	0.610	0.549	NATI	61.9	38.9	0.561	0.560
AMZN	61.9	49.3	0.482	0.605	NDAQ	66.4	54.9	0.597	0.404
BA	61.6	76.4	0.650	0.857	NFLX	68.2	84.7	0.755	0.917
BB	70.3	49.3	0.732	0.618	NOK	59.2	47.9	0.585	0.510
CAJ	59.5	49.3	0.574	0.592	NVDA	76.3	98.6	0.851	0.993
CVX	61.9	55.6	0.557	0.584	ORCL	64.6	50.7	0.607	0.632
DAL	68.5	54.2	0.677	0.431	PEP	66.1	50.0	0.641	0.532
DOX	67.6	45.1	0.662	0.586	PRU	58.3	53.5	0.570	0.599
DVMT	61.6	48.6	0.508	0.507	PTR	61.0	50.0	0.561	0.446
F	63.1	56.3	0.602	0.512	QCOM	64.9	51.4	0.626	0.533
FB	61.9	47.2	0.567	0.638	RDS-B	65.5	53.5	0.602	0.362
GE	58.9	44.4	0.519	0.245	S	66.4	60.4	0.687	0.565
GOOGL	67.0	45.1	0.628	0.568	SGTZY	67.9	54.9	0.616	0.463
HMC	68.2	54.2	0.639	0.560	SNE	57.4	50.0	0.496	0.636
HP	66.1	50.0	0.641	0.486	STX	59.8	53.5	0.518	0.496
HPQ	83.2	97.9	0.901	0.989	T	62.5	54.2	0.603	0.154
IBM	62.8	47.9	0.635	0.272	TSLA	65.8	47.9	0.672	0.566
INTC	64.3	57.6	0.607	0.358	V	71.8	95.8	0.795	0.979
K	65.5	56.9	0.625	0.404	VZ	59.2	46.5	0.547	0.374
KINS	64.3	47.9	0.629	0.576	WDC	61.3	50.7	0.608	0.553
LOGI	67.9	41.0	0.640	0.536	XOM	63.1	59.7	0.599	0.525

From the results in Table 2, we observe that the testing accuracy for individual stocks varies between 38.9% and 98.6%. The F1 Score shows a similar trend as the classification accuracy, having a large variance in F1 Score across 50 stocks. Hence, it is evident that classifier does not show a consistent classification performance for stock price movements.

Although the classification performance are generally poor, the classifier does well in some stocks. The Fusion Learning Method achieves high testing accuracy and F1 Score for some datsets such as MSFT and NVDA, which indicates a good potential for classifying the movement of stock market.

With these observations, it is important that the classifier chooses a only subset of stocks to predict such that its potential is maximized and the errors are minimized. The classification performance of our proposed method and the effect of adjusting the window length are presented in Tables 3 and 4.

Table 3. Classification accuracy on test set

Top ranked no. of stocks	Window length (days)					
	50	100	150	200	250	300
50	56.8	56.8	56.8	56.8	56.8	56.8
25	62.6	**63.4**	63.3	62.9	62.5	62.7
15	69.6	**70.4**	69.4	69.2	68.8	68.3
10	77.3	**79.0**	78.0	76.3	75.5	74.4
5	90.0	90.4	**90.8**	89.6	88.5	88.9
4	93.4	**93.9**	**93.9**	93.2	92.7	92.9
3	97.5	97.5	97.5	97.5	97.5	97.5
2	97.9	98.3	98.3	98.3	98.3	98.3
1	97.2	97.9	97.9	97.9	97.9	97.9

Bold values indicate optimal window length for the number of stocks selected.

Table 4. Classification F1 score on test set

Top ranked no. of stocks	Window length (days)					
	50	100	150	200	250	300
50	0.632	0.632	0.632	0.632	0.632	0.632
25	0.702	**0.704**	**0.704**	0.699	0.697	0.702
15	0.788	**0.791**	0.777	0.778	0.770	0.766
10	0.861	**0.872**	0.864	0.847	0.837	0.832
5	0.946	0.949	**0.951**	0.944	0.938	0.940
4	0.966	**0.969**	**0.969**	0.965	0.962	0.962
3	0.987	0.987	0.987	0.987	0.987	0.987
2	0.989	0.991	0.991	0.991	0.991	0.991
1	0.986	0.989	0.989	0.989	0.989	0.989

Bold values indicate optimal window length for the number of sets selected.

In Tables 3 and 4, we observed that our approach works well in prioritizing on easy-to-classify problems in the test set. Both classification accuracy and F1 Score improves when less number of stocks are selected. In addition, there are significant improvements when only 5 (10%) or less the number of stocks are selected. Selecting the top 2 stocks results in the best classification performance.

In addition, it shows that the window length does not show a significant effect on classification performance. Accuracy and F1 Score remains largely the same with different window lengths when the top {3, 2, 1} stocks are selected. Setting the window length to 150 days would achieve the best classification result as it produces the best results when 5 of less stocks are selected, where it achieves significantly better results as compared to other cases which uses 10 or more stocks.

5   Conclusion and Future Work

In this paper, we proposed a new approach which allows the ensemble learning model to further improve its effectiveness in predicting the direction of stock market movements. The proposed approach involves training and evaluating the classification models on many stocks and prioritizing on the easy-to-classify stocks to maximize the potential of state-of-the-art classifiers. Experiments on 50 United States stocks showed that the classification performance improves when the classifier prioritizes on a smaller subset of stock.

This work can be extended in some ways. Firstly, more efficient models can be experimented in place of the ensemble learning method used in this work. For example, using an incremental ensemble learning model which updates itself when new samples becomes available. The experiment can be expanded to include more stocks and longer historical data. In addition, work can also be extended to other applications such as foreign currency exchange rates and commodity prices. Lastly, the experiment can be improved by applying cross-validation on training-testing splits and including other performance matrices such as AUC to give a more complete review on the learning performance of the algorithm.

References

1. Barak, S., Arjmand, A., Ortobelli, S.: Fusion of multiple diverse predictors in stock market. Inf. Fusion **36**, 90–102 (2017). https://doi.org/10.1016/j.inffus.2016.11.006
2. Breiman, L.: Stacked regressions. Mach. Learn. **24**(1), 49–64 (1996). https://doi.org/10.1007/BF00117832
3. Breiman, L.: Random forests. Mach. Learn. **45**(1), 5–32 (2001). https://doi.org/10.1023/A:1010933404324
4. Chen, Y., Feng, M.Q.: A technique to improve the empirical mode decomposition in the Hilbert-Huang transform. Earthq. Eng. Eng. Vib. **2**(1), 75–85 (2003). https://doi.org/10.1007/BF02857540
5. Cortes, C., Vapnik, V.: Support-vector networks. Mach. Learn. **20**(3), 273–297 (1995). https://doi.org/10.1007/BF00994018
6. Darbellay, G.A., Slama, M.: Forecasting the short-term demand for electricity: do neural networks stand a better chance? Int. J. Forecast. **16**(1), 71–83 (2000). https://doi.org/10.1016/S0169-2070(99)00045-X
7. Ding, X., Zhang, Y., Liu, T., Duan, J.: Deep learning for event-driven stock prediction. In: Ijcai, pp. 2327–2333 (2015)
8. Drucker, H., Burges, C.J., Kaufman, L., Smola, A.J., Vapnik, V.: Support vector regression machines. In: Advances in Neural Information Processing Systems, pp. 155–161 (1997)
9. Grossmann, A., Morlet, J.: Decomposition of hardy functions into square integrable wavelets of constant shape. SIAM J. Math. Anal. **15**(4), 723–736 (1984). https://doi.org/10.1137/0515056
10. Guan, C., Luh, P.B., Michel, L.D., Wang, Y., Friedland, P.B.: Very short-term load forecasting: wavelet neural networks with data pre-filtering. IEEE Trans. Power Syst. **28**(1), 30–41 (2013). https://doi.org/10.1109/TPWRS.2012.2197639

11. He, K., Zha, R., Wu, J., Lai, K.K.: Multivariate EMD-based modeling and forecasting of crude oil price. Sustainability **8**(4) (2016). https://doi.org/10.3390/su8040387
12. Hinton, G.E., Salakhutdinov, R.R.: Reducing the dimensionality of data with neural networks. Science **313**(5786), 504–507 (2006). https://doi.org/10.1126/science.1127647
13. Hochreiter, S., Schmidhuber, J.: Long short-term memory. Neural Comput. **9**(8), 1735–1780 (1997). https://doi.org/10.1162/neco.1997.9.8.1735
14. Holt, C.C.: Forecasting seasonals and trends by exponentially weighted moving averages. Int. J. Forecast. **20**(1), 5–10 (2004). https://doi.org/10.1016/j.ijforecast.2003.09.015
15. Hooshmand, R.A., Amooshahi, H., Parastegari, M.: A hybrid intelligent algorithm based short-term load forecasting approach. Int. J. Electr. Power Energy Syst. **45**(1), 313–324 (2013). https://doi.org/10.1016/j.ijepes.2012.09.002
16. Huang, N.E., et al.: The empirical mode decomposition and the Hilbert spectrum for nonlinear and non-stationary time series analysis. In: Proceedings of the Royal Society of London A: Mathematical, Physical and Engineering Sciences, vol. 454, pp. 903–995. The Royal Society (1998)
17. Janacek, G.: Time series analysis forecasting and control. J. Time Ser. Anal. **31**(4), 303–303 (2010). https://doi.org/10.1111/j.1467-9892.2009.00643.x
18. Kara, Y., Boyacioglu, M.A., Baykan, O.K.: Predicting direction of stock price index movement using artificial neural networks and support vector machines: the sample of the Istanbul stock exchange. Expert Syst. Appl. **38**(5), 5311–5319 (2011). https://doi.org/10.1016/j.eswa.2010.10.027
19. Kiplangat, D.C., Asokan, K., Kumar, K.S.: Improved week-ahead predictions of wind speed using simple linear models with wavelet decomposition. Renew. Energy **93**, 38–44 (2016). https://doi.org/10.1016/j.renene.2016.02.054
20. Krizhevsky, A., Sutskever, I., Hinton, G.E.: ImageNet classification with deep convolutional neural networks. In: Pereira, F., Burges, C.J.C., Bottou, L., Weinberger, K.Q. (eds.) Advances in Neural Information Processing Systems, vol. 25, pp. 1097–1105. Curran Associates, Inc. (2012)
21. Pao, Y.H., Park, G.H., Sobajic, D.J.: Learning and generalization characteristics of the random vector functional-link net. Neurocomputing **6**(2), 163–180 (1994). https://doi.org/10.1016/0925-2312(94)90053-1. Backpropagation, Part IV
22. Pao, Y.H., Phillips, S.M., Sobajic, D.J.: Neural-net computing and the intelligent control of systems. Int. J. Control **56**(2), 263–289 (1992). https://doi.org/10.1080/00207179208934315
23. Papalexopoulos, A.D., Hesterberg, T.C.: A regression-based approach to short-term system load forecasting. IEEE Trans. Power Syst. **5**(4), 1535–1547 (1990). https://doi.org/10.1109/59.99410
24. Percival, D.B., Walden, A.T.: Wavelet Methods for Time Series Analysis, vol. 4. Cambridge University Press, Cambridge (2006)
25. Percival, D.B., Wang, M., Overland, J.E.: An introduction to wavelet analysis with applications to vegetation time series. Commun. Ecol. **5**(1), 19–30 (2004)
26. Qiu, X., Zhang, L., Ren, Y., Suganthan, P.N., Amaratunga, G.: Ensemble deep learning for regression and time series forecasting. In: 2014 IEEE Symposium on Computational Intelligence in Ensemble Learning (CIEL), pp. 1–6, December 2014. https://doi.org/10.1109/CIEL.2014.7015739
27. Qiu, X., Ren, Y., Suganthan, P.N., Amaratunga, G.A.: Empirical mode decomposition based ensemble deep learning for load demand time series forecasting. Appl. Soft Comput. **54**, 246–255 (2017). https://doi.org/10.1016/j.asoc.2017.01.015

28. Qiu, X., Suganthan, P.N., Amaratunga, G.A.J.: Fusion of multiple indicators with ensemble incremental learning techniques for stock price forecasting. J. Bank. Financ. Technol. (2019). https://doi.org/10.1007/s42786-018-00006-2

29. Qiu, X., Suganthan, P.N., Amaratunga, G.A.: Ensemble incremental learning random vector functional link network for short-term electric load forecasting. Knowl.-Based Syst. **145**, 182–196 (2018). https://doi.org/10.1016/j.knosys.2018.01.015

30. Ren, Y., Suganthan, P., Srikanth, N., Amaratunga, G.: Random vector functional link network for short-term electricity load demand forecasting. Inf. Sci. **367–368**, 1078–1093 (2016). https://doi.org/10.1016/j.ins.2015.11.039

31. Ren, Y., Zhang, L., Suganthan, P.N.: Ensemble classification and regression-recent developments, applications and future directions. IEEE Comp. Int. Mag. **11**(1), 41–53 (2016)

32. Schmidt, W.F., Kraaijveld, M.A., Duin, R.P.W.: Feedforward neural networks with random weights. In: 11th IAPR International Conference on Pattern Recognition, Proceedings. Conference B: Pattern Recognition Methodology and Systems, vol. II, pp. 1–4, August 1992. https://doi.org/10.1109/ICPR.1992.201708

33. Sokolova, M., Lapalme, G.: A systematic analysis of performance measures for classification tasks. Inf. Process. Manag. **45**(4), 427–437 (2009). https://doi.org/10.1016/j.ipm.2009.03.002

34. Wu, Z., Huang, N.E.: Ensemble empirical mode decomposition: a noise-assisted data analysis method. Adv. Adapt. Data Anal. **01**(01), 1–41 (2009). https://doi.org/10.1142/S1793536909000047

35. Zhang, L., Suganthan, P.: A comprehensive evaluation of random vector functional link networks. Inf. Sci. **367–368**, 1094–1105 (2016). https://doi.org/10.1016/j.ins.2015.09.025

A Neural Net Based Prediction of Sound Pressure Level for the Design of the Aerofoil

Palash Pal[1], Rituparna Datta[2(✉)], Deepak Rajbansi[3], and Aviv Segev[2]

[1] University Institute of Technology, Burdwan University, Bardhaman, India
pal.palash.ml.research@gmail.com
[2] Department of Computer Science,
University of South Alabama, Mobile, AL, USA
{rdatta,segev}@southalabama.edu
[3] Design and Validation, QuEST Global Engineering Services Private Limited,
Bengaluru, India
deepak.rajbansi@quest-global.com

Abstract. Aerofoil self-noise can affect the performance of the overall system. One of the main goals of aircraft design is to create an aerofoil with minimum weight, cost, and self-noise, satisfying all design requirements from the physical and the functional requirements. Aerofoil self-noise refers to the noise produced by the interaction between an aerofoil and its boundary layer. This paper describes how the prediction of the self-noise of an aerofoil at the early stage of the design phase can help select the best design of the aerofoil, which in turn reduces the lead time as the design process becomes more robust with respect to cost effectiveness. In the present work, the prediction of the self-noise of the aerofoil is addressed using Neural Networks (NN). Different architectures are used along with various proportions of training and testing to select the best architecture and best training-testing ratio. The results from NN is compared with linear, quadratic, and cubic polynomial regression. Thereafter, Principal Component Analysis (PCA) is integrated with NN for further improvement of prediction results. Our experimental results indicate that neural networks outperform regression. Moreover, PCA integrated with NN outperforms even the best neural network result.

Keywords: Neural network · Back propagation · Design of aerofoil · Principal Component Analysis (PCA) · Regression

1 Introduction

Aerofoil is the cross-sectional shape of a wing, blade, or sail [6]. An aerofoil shaped body moving through fluid produces an aerodynamic force [9,18]. The component of this force perpendicular to the direction of motion is called lift. The component parallel to the direction of motion is called drag [13,15,19].

© Springer Nature Switzerland AG 2020
A. Zamuda et al. (Eds.): SEMCCO 2019/FANCCO 2019, CCIS 1092, pp. 105–112, 2020.
https://doi.org/10.1007/978-3-030-37838-7_10

Aerofoils are used in aircraft as wings to produce lift or as propeller blades to produce thrust. While designing aircraft, design requirements for wings or propeller blades are fixed initially. Many parameters are taken into consideration [4,10]. In case of the design of the wing, parameters considered are lift, drag, Critical Mach Number, angle of attack, coefficient of lift, etc. For the design of the blade parameters considered are lift, drag, pressure difference, flow rate, power generated, efficiency, angle of attack, etc. After freezing the design requirements, we need to choose suitable aerofoils that fulfil all the design requirements. Some research on aerofoil design optimisation is available in [7,12]. Self-noise is the noise produced by one's own body. Self-noise of an aerofoil is generated by its boundary layer turbulence interacting with the trailing edge of the blades or wings. Aerofoil self-noise reduces blade efficiency and increases flutter in case of wings, thus affecting system performance and integrity [3,17]. It is important to predict aerofoil self-noise before finalizing the design of wings or blades to prevent any unwanted overall behaviour of the system as a whole. Another study by Vathylakis et al. [3] performed a study to determine different variables which affect the self-noise reduction of an aerofoil. Chong et al. [17] performed an experimental study to reduce aerofoil self-noise.

There is no standard method available to predict aerofoil self-noise as the relationship between the input parameters and the output self-noise is random in nature. In our present experiment, an effort is made to model input parameters against output using different machine learning methods. We applied linear, quadratic, and cubic regression; neural network model [2,16], and neural network with PCA [5,8] to predict aerofoil self-noise.

The organisation of the rest of the present paper is as follows. Section 2 discusses the description of the self-noise dataset, provides the details of the method, and the simulation results. The future scope along with the conclusion is in Sect. 3.

2   Proposed Method and Simulation Results

2.1   Self-noise Dataset Description

An existing open source dataset [1] is used for the present study to predict the self-noise of the aerofoil where the output is sound pressure level in decibel with five inputs. There are millions of aerofoils available which are pre-designed. Among them, more than one design may satisfy specific design requirements. Our aim is to select the most suitable aerofoil from available designs which satisfies all the design requirements and produces minimum noise (scaled sound pressure level). There is no standard mathematical formulation to determine the noise level of an aerofoil from given input data. We used Neural Network to achieve that goal. Some earlier studies that used the same dataset can be found in [11,14]. Our input parameters are:

1. Frequency in Hertz
2. Angle of attack in degree

3. Chord length in meter
4. Free stream velocity in meter per second
5. Suction side displacement thickness in meter

Our output is scaled sound pressure level in decibels. In this experiment, our effort is to predict aerofoil noise to select the aerofoil with minimum noise that satisfies all the design requirements.

2.2 Proposed Method

The neural network model is suitable when there is a nonlinear and random relationship between input and output variables. From the scatterplot matrix in Fig. 1, it is clear that different input parameters in our present experiment have a nonlinear and random relationship to the output. So, we used a neural network to predict the output against input combinations. The prediction problem is solved using multilayer feed forward back propagation network. The proposed Neural Network (NN) is considered to have five inputs (frequency in Hertz, angle of attack in degree, chord length in meter, free stream velocity in meter per second, suction side displacement thickness in meter).

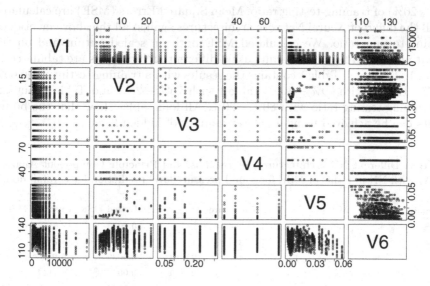

Fig. 1. Scatter plot matrix of input and output variables for visual representation.

Regression analysis is suitable to model the relationship between independent variables and dependent variables when the dependent variable is continuous in nature. In our present experiment, the dependent variable is aerofoil self-noise, which is a continuous variable in nature. So, we applied regression to model the relationship between the input and output parameters. Principal component analysis is used to transform input variables into a set of linearly uncorrelated variables. It is also a process of dimensionality reduction where less important components can be ignored while modelling the principal components against the output. In our experiment, we ignored the last component as its proportion of variance is negligible.

2.3   Simulation Results

In this section we explain the simulation results. Figure 1 presents a scatter plot matrix between five input variables with the output variable.

This plot gives an insight about visual representation of the data along with correlation between variables. The figure of the scatter plot matrix clearly shows a random relationship between the different parameters. As a result, a neural network is appropriate to handle the prediction problem.

In our experiment, first of all we shuffled the data rows. Thereafter, we scaled the whole data set using Max-Min scaling to have all the attribute values lying between zero to one. We ignored the data cleaning step as data already was cleaned. We used different Neural Network architectures to compare the results. Values represent number of neurons in each layer. We also tested all the network architectures against different proportion (60% – 40%, 70% – 30%, 75% – 25%, 80% – 20%) of training-testing ratio. Mean Squared Errors (MSE) are calculated for all the different Neural Network architectures against all the above mentioned training-testing ratio. We calculated MSE both on scaled output and on the output after unscaling. The results using different architectures are tabulated in Tables 1, 2, 3 and 4. Table 1 contains the results with a training-testing ratio 60% – 40%, Table 2 with 70% – 30%, Table 3 with 75% – 25% and Table 4 contains the training-testing ratio with 80% – 20%. All four tables show that the better results can be achieved using a greater number of hidden layers.

Table 1. Different NN architectures, regressions, and Principle Component Analysis (PCA) with NN in terms of MSE for sound pressure level prediction for aerofoil design

Training – Testing	Method	Architecture	Scaled MSE	Unscaled MSE
60% – 40%	Neural Network	5 — 3 — 1	0.0073	12.8172
		5 — 3 — 2 — 1	0.0097	13.7662
		5 — 4 — 3 — 2 — 1	0.0032	**5.2197**
		5 — 5 — 3 — 2 — 1	0.0043	6.4114
		5 — 4 — 4 — 3 — 2 — 1	0.0042	6.2937
	Regression	Linear		25.61517
		Quadratic		21.39708
		Cubic		18.93659
	PCA with NN	4 — 6 — 4 — 2 — 1	0.0037	5.5179
		4 — 6 — 4 — 4 — 2 — 1	0.0021	**4.0452**
		4 — 6 — 4 — 3 — 2 — 1	0.0025	4.6141
		4 — 6 — 5 — 4 — 2 — 1	0.0033	5.7024

Table 2. Different NN architectures, regressions, and Principle Component Analysis (PCA) with NN in terms of MSE for sound pressure level prediction for aerofoil design

Training – Testing	Method	Architecture	Scaled MSE	Unscaled MSE
70% – 30%	Neural Network	5 — 3 — 1	0.0062	10.7369
		5 — 3 — 2 — 1	0.0055	8.1798
		5 — 4 — 3 — 2 — 1	0.0039	6.6042
		5 — 5 — 3 — 2 — 1	0.0031	**5.0883**
		5 — 4 — 4 — 3 — 2 — 1	0.0033	5.3859
	Regression	Linear		25.53724
		Quadratic		25.11363
		Cubic		20.08023
	PCA with NN	4 — 6 — 4 — 2 — 1	0.0037	5.4129
		4 — 6 — 4 — 4 — 2 — 1	0.0027	4.7193
		4 — 6 — 4 — 3 — 2 — 1	0.0028	**4.1058**
		4 — 6 — 5 — 4 — 2 — 1	0.0037	5.2964

Table 3. Different NN architectures, regressions, and Principle Component Analysis (PCA) with NN in terms of MSE for sound pressure level prediction for aerofoil design

Training – Testing	Method	Architecture	Scaled MSE	Unscaled MSE
75% – 25%	Neural Network	5 — 3 — 1	0.0056	10.5351
		5 — 3 — 2 — 1	0.0054	10.2200
		5 — 4 — 3 — 2 — 1	0.0044	6.7297
		5 — 5 — 3 — 2 — 1	0.0036	5.3320
		5 — 4 — 4 — 3 — 2 — 1	0.0028	**4.3825**
	Regression	Linear		23.78459
		Quadratic		19.87766
		Cubic		18.71646
	PCA with NN	4 — 6 — 4 — 2 — 1	0.0024	3.6856
		4 — 6 — 4 — 4 — 2 — 1	0.0024	3.6856
		4 — 6 — 4 — 3 — 2 — 1	0.0034	5.1138
		4 — 6 — 5 — 4 — 2 — 1	0.0021	**3.2747**

We also modelled the same dataset using regression (linear, quadratic and cubic). MSE in all of these cases are calculated. The results of regressions are also tabulated in Tables 1, 2, 3 and 4 along with NN results. From the results in Tables 1, 2, 3 and 4, it is clear that Neural Network outperforms regression.

According to our study, the best configuration is with 5-4-4-3-2-1 network architecture with training-testing ratio as 75% – 25%. The worst neural network performance is better than the best regression model.

Finally, we used principal component analysis to find out the principal components and the results are tabulated in Table 5. From the table it is clear that the first four components are important as the cumulative proportion of the first four components is as high as 96%. Figures 3 and 2 present the importance of various principal components. After using the first four principal components as

Table 4. Different NN architectures, regressions, and Principle Component Analysis (PCA) with NN in terms of MSE for sound pressure level prediction for aerofoil design

Training – Testing	Method	Architecture	Scaled MSE	Unscaled MSE
80% – 20%	Neural Network	5 — 3 — 1	0.0070	12.3160
		5 — 3 — 2 — 1	0.0052	7.8561
		5 — 4 — 3 — 2 — 1	0.0055	9.4192
		5 — 5 — 3 — 2 — 1	0.0042	7.9350
		5 — 4 — 4 — 3 — 2 — 1	0.0032	**4.9800**
	Regression	Linear		23.87245
		Quadratic		19.85895
		Cubic		22.41482
	PCA with NN	4 — 6 — 4 — 2 — 1	0.0023	**3.9487**
		4 — 6 — 4 — 4 — 2 — 1	Non convergence	
		4 — 6 — 4 — 3 — 2 — 1	0.0035	5.0018
		4 — 6 — 5 — 4 — 2 — 1	0.0024	4.0019

Table 5. Importance of components

	PC1	PC2	PC3	PC4	PC5
Standard deviation	1.452374	1.060189	0.9573568	0.8220337	0.4175374
Proportion of variance	0.421880	0.224800	0.1833100	0.1351500	0.0348700
Cumulative proportion	0.421880	0.646680	0.8299800	0.9651300	1.0000000

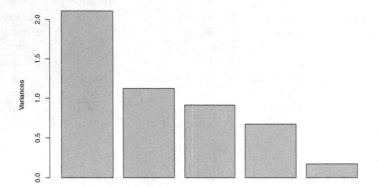

Fig. 2. Variance of the five principal components.

input to our neural network, results are improved. Among four different architectures against four training-testing combinations for each, in nine cases the PCA neural network hybrid outperformed the best neural network result.

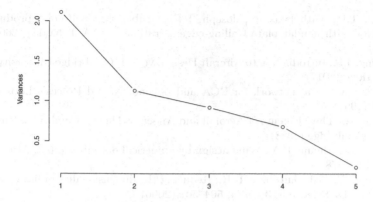

Fig. 3. Scree plot to show the variance of different principal components.

3 Conclusion

The self-noise of an aerofoil may lead to structural failure of a wing or a fan blade. It drastically reduces the fan blade efficiency, and further it affects the environment by producing aerodynamic noise. Predicting aerofoil self-noise is a very important task as the overall behaviour of a system where blades or wings are used depends on it. In this paper we made an effort to apply different machine learning techniques. From our experiment it is obvious that a neural net performs much better than regression. Among regression techniques, quadratic regression performs better than linear and in most of the cases, cubic performs better than quadratic. After applying PCA and using only the important components in the neural network to predict aerofoil self-noise, performance becomes better than even the best neural network result. Although in our experiment we obtained significant results, there is scope for further experimentation. Regression with PCA can be used to compare results with ordinary regression. While applying PCA on the neural network, we ignored only the last component as the first four components have cumulative proportion of 97%. For further experimentation, the last two components can be ignored as the first three components have 83% of cumulative proportion. After that, results can be compared to draw further conclusions. In our present experiment, as the best result produces output with really high accuracy, the proposed method could be implemented practically to predict aerofoil self-noise. Overall this methodology has many advantages over conventional prediction methodology as this model can be used for any sets of aerofoils whether for a wing design or a rotor blade design.

References

1. Brooks, T.F., Pope, D.S., Marcolini, M.A.: Airfoil self-noise and prediction. Technical report, NASA (1989)
2. Cao, Y., Yuan, K., Li, G.: Effects of ice geometry on airfoil performance using neural networks prediction. Aircr. Eng. Aerosp. Technol. **83**(5), 266–274 (2011)

3. Chong, T.P., Vathylakis, A., Joseph, P.F., Gruber, M.: Self-noise produced by an airfoil with nonflat plate trailing-edge serrations. AIAA J. **51**(11), 2665–2677 (2013)
4. Fielding, J.P.: Introduction to Aircraft Design, vol. 11. Cambridge University Press, Cambridge (2017)
5. Fyfe, C.: A neural network for PCA and beyond. Neural Process. Lett. **6**(1–2), 33–41 (1997)
6. Glauert, H.: The Elements of Aerofoil and Airscrew Theory. Cambridge University Press, Cambridge (1983)
7. Hicks, R.M., Henne, P.A.: Wing design by numerical optimization. J. Aircr. **15**(7), 407–412 (1978)
8. Hinton, G.E., Salakhutdinov, R.R.: Reducing the dimensionality of data with neural networks. Science **313**(5786), 504–507 (2006)
9. Kesel, A.B.: Aerodynamic characteristics of dragonfly wing sections compared with technical aerofoils. J. Exp. Biol. **203**(20), 3125–3135 (2000)
10. Kundu, A.K.: Aircraft Design, vol. 27. Cambridge University Press, Cambridge (2010)
11. Lau, K., et al.: A neural networks approach for aerofoil noise prediction. Master thesis (2006)
12. LeGresley, P., Alonso, J.: Airfoil design optimization using reduced order models based on proper orthogonal decomposition. In: Fluids 2000 Conference and Exhibit, p. 2545 (2000)
13. Lock, R.: The prediction of the drag of aerofoils and wings at high subsonic speeds. Aeronaut. J. **90**(896), 207–226 (1986)
14. Lopez, R., Balsa-Canto, E., Oñate, E.: Neural networks for variational problems in engineering. Int. J. Numer. Meth. Eng. **75**(11), 1341–1360 (2008)
15. Sanaye, S., Hassanzadeh, A.: Multi-objective optimization of airfoil shape for efficiency improvement and noise reduction in small wind turbines. J. Renew. Sustain. Energy **6**(5), 053105 (2014)
16. Srivastava, S., Roy, A.K., Kumar, K.: Design analysis of mixed flow pump impeller blades using ansys and prediction of its parameters using artificial neural network. Procedia Eng. **97**, 2022–2031 (2014)
17. Vathylakis, A., Paruchuri, C.C., Chong, T.P., Joseph, P.: Sensitivity of aerofoil self-noise reductions to serration flap angles. In: 22nd AIAA/CEAS Aeroacoustics Conference, p. 2837 (2016)
18. Withers, P.C.: An aerodynamic analysis of bird wings as fixed aerofoils. J. Exp. Biol. **90**(1), 143–162 (1981)
19. Wortmann, F.: Progress in the design of low drag aerofoils. In: Boundary Layer and Flow Control, pp. 748–770. Elsevier (1961)

Competition of Strategies
in jSO Algorithm

Petr Bujok[✉][iD]

University of Ostrava, 30. dubna 22, 70200 Ostrava, Czech Republic
petr.bujok@osu.cz

Abstract. A newly proposed variant of the efficient jSO algorithm employing competition of eight strategies (cSO) is proposed. The main idea is to select the most proper strategy and adapt the setting to each solved problem. One more mutation variant and one more type of crossover are added to jSO, and moreover, the popular mechanism of Eigen coordinate system is applied. All eight strategies compete to be used in the next generations based on the successes in previous generations. The proposed cSO method has more wins over jSO significantly in more real-world problems than fails. The original jSO strategy is never the most frequently used strategy, compared with other newly employed strategies.

Keywords: Differential evolution · jSO · Strategy · Competition · Real-world problems

1   Differential Evolution

The Differential Evolution (DE) algorithm is a popular optimisation technique proposed in 1995 by Storn and Price [11]. Against the simplicity of this method, DE is very efficient compared with other optimisation algorithms. During more than twenty years, a lot of enhanced DE variants with increasing efficiency have been introduced [7]. The popularity of the differential evolution algorithm is also evident from many competitions and real-applications [10,16]. Similarly, ensemble strategies are widely used in various population-based algorithms [15].

One of the non-negligible milestones came in 2014 when a popular adaptive L-SHADE variant [12] was introduced. In this method, several new elements were joined together and provided the best results in competition CEC 2014. Then, after three generations of this algorithm, a very efficient jSO variant [1] was proposed and it took overall the second place in CEC 2017 competition.

The main aim of this paper is to increase the efficiency of the successful jSO algorithm using a competition of several strategies. The inspiration was taken from the competition of four different strategies applied to the preceding SHADE variant, which brings significantly better results compared with the original algorithm [6]. A newly proposed version of jSO is compared with the original one when solving real-world problems.

© Springer Nature Switzerland AG 2020
A. Zamuda et al. (Eds.): SEMCCO 2019/FANCCO 2019, CCIS 1092, pp. 113–121, 2020.
https://doi.org/10.1007/978-3-030-37838-7_11

The remaining text of the paper is formed as follows. Section 1.1 provides a brief description of the original jSO algorithm. A competition of strategies in jSO is described in Sect. 2. Experimental configuration of problems and methods are described in Sect. 3. Results and statistical analysis on real-world problems are presented in Sect. 4. Section 5 concludes the paper and highlights some findings.

1.1 Adaptive DE Variant – jSO

As said, one of the latest best performing adaptive DE variant which takes a very good position at international competitions is jSO [1]. The jSO algorithm uses self-adaptive approach of parameters. When the user of jSO uses the recommended initial setting, this method is parameter-free.

Beside popular current-to-$p$best mutation strategy, jSO uses also an archive A of old good parent individuals. Moreover, circle memories to adaptation of the control parameters F (μ_F) and CR (μ_{CR}) are employed. The jSO algorithm uses initial values $\mu_{CR} = 0.8$ and $\mu_F = 0.3$. When the last Hth position is selected, the same values $\mu_{CR} = \mu_F = 0.9$ are used. jSO uses a linear reduction of the population size, introduced in L-SHADE. The comprehensive description of jSO is in the original paper [1].

Although the control parameters of the jSO variant are tuned properly, a cooperating model of several jSO algorithms provides good performance in some real-world problems [2]. In this paper, eight different strategies compete to be used in the reproductive process.

2 Competition of Strategies in jSO

The original jSO algorithm uses only one combination of mutation (current-to-$p$best) and crossover (binomial). In a comprehensive study of various DE strategies [5] was shown, that mutation variant *randrl/1* provides very good performance. Besides this, an efficient exponential crossover achieves very good results when applied to state-of-the-art DE variants and solutions of real-world problems [4]. The four combinations of two mutations current-to-$p$best and randrl/1, along with two types of crossover (binomial and exponential), compete to be used in a population of SHADE [6].

Motivated by previously mentioned experiments, the efficient jSO algorithm is enhanced by two different mutation variants combined with two kinds of crossover. These four strategies are used in a standard form, and another four-some uses the same combinations extended by Eigenvector coordinate system for a crossover operation. Therefore, eight strategies compete to be used in the reproduction of a population of jSO, where a more successful strategy is preferred in the next generations, and vice versa. This newly proposed competitive variant of jSO algorithm is simply called *cSO*.

2.1   Mutation Variants

The original jSO algorithm inherits a popular current-to-$p$best mutation variant (1), where current point x_i, randomly selected point from population x_{r1}, randomly selected point from union of population and archive x_{r2}, and randomly selected point from p best points of population x_{pbest} are used.

$$u_i = x_i + F_i\left(x_{pbest} - x_i\right) + F_i\left(x_{r1} - x_{r2}\right).\tag{1}$$

Although the original jSO algorithm provides very good results, cSO also uses the second well-performing mutation variant randrl/1:

$$u_i = x_{r1} + F\left(x_{r2} - x_{r3}\right)\tag{2}$$

where the point x_{r1} is the tournament best among x_{r1}, x_{r2}, and x_{r3}, i.e. $f(x_{r1}) \leq f(x_j)$, $j = 2, 3$.

2.2   Type of Crossover

The original jSO variant uses the popular binomial crossover (3). In the proposed cSO algorithm, a very efficient exponential crossover variant is also employed (4).

$$y_{i,j} = \begin{cases} u_{i,j} \; if\, rand_j(0,1) \leq CR \; or \; j = rand(1, D) \\ x_{i,j} \; otherwise, \end{cases}\tag{3}$$

where $rand_j$ is a random number from $(0, 1)$, and j is a random index from $(1, D)$.

$$y_{i,j} = \begin{cases} u_{i,j} \; for \; j = \langle n\rangle_D, \; \langle n+1\rangle_D, \ldots, \langle n + L - 1\rangle_D \\ x_{i,j} \; otherwise, \end{cases}\tag{4}$$

where the brackets $\langle\rangle_D$ represent the modulo function with modulus D. The starting position of crossover (n) is selected randomly from $\{1, \ldots, D\}$, and L consecutive elements are selected from the mutant vector u_i.

2.3   Eigenvector Coordinate System

In 2014, the covariance-based Eigenvector coordinate system for a crossover operation in DE (CoBiDE) was proposed [14]. The aim was a higher efficiency in problems with highly correlated coordinates. This crossover variant was used in a latter study where it achieved good results [3]. A covariance matrix C from a part of individuals (ps) is used. The matrix C is divided into two components, Eigenvectors B and Eigenvalues D:

$$C = BD^2B^T\tag{5}$$

A mutation point u_i and the current point x_i are combined in the Eigen coordinate system, with probability pb:

$$x_i' = B^T x_i \quad \text{and} \quad u_i' = B^T u_i. \tag{6}$$

Then, newly developed trial vector y_i' is transformed from the Eigen coordinate system back to a standard coordinate system.

$$y_i = B y_i' \tag{7}$$

This transformation is performed only with a probability pb, in other cases, the crossover operation in a standard coordinate system is preferred.

2.4   Competition Mechanism

In cSO, four strategies use a standard coordinate system and four same strategies are enhanced by the Eigenvector approach, i.e. trial points of the whole population are generated using (6) and (7). All eight strategies compete to be used in next generations. The competition mechanism was introduced in [13]. After each generation, the probability to be used q_k for each of K strategies is updated in dependency on the success of the currently used strategy in previous generations (8). A more successful strategy is preferred frequently than a strategy which is not able to generate successful individuals.

$$q_k = \frac{n_k + n_0}{\sum_{j=1}^{K} (n_j + n_0)}, \tag{8}$$

where q_k is computed probability to be used of the kth DE strategy, n_k is the count of the kth strategy successes, $n_0 = 2$ prevents a huge change in q_k. The values of probabilities are reset to the initial values if any q_k value decreases below a input parameter δ, $\delta > 0$.

3   Experimental Settings

A test suite of 22 real-world problems used in the CEC 2011 for the Special Session on Real-Parameter Numerical optimisation [8] is applied in this comparison. The main differences between functions of this set are in computational complexity, and the dimensionality of the search area is from $D = 1$ to $D = 240$. Each method performs 25 independent runs on each problem. The run of the algorithm is stopped after the prescribed number of function evaluation $MaxFES$ $= 150000$. The results of the algorithms are also recorded in one third and two-thirds of $MaxFES$. The individual with the smallest function value represents the solution of the problem.

Both compared algorithms use a linear population-size reduction with initial value $N_{\text{init}} = \text{round}(25 \times \log(D) \times \sqrt{D})$. For problems with a low dimension level $D < 6$, initial population size is computed from value $D = 6$. The number of

Table 1. Results of comparison of new cSO algorithm with jSO from Wilcoxon rank-sum test.

F	D	Best	Worst	Median	Mean	SD	jSO (med.)	p
T01	6	0	12.5353	0	1.92359	4.50575	0	≈
T02	30	−28.4225	−26.3287	**−27.1149**	−26.984	0.528249	−26.295	***
T03	1	1.15E−05	1.15E−05	1.15E−05	1.15E−05	5.19E−21	1.15E−05	≈
T04	1	13.7708	20.9574	14.3291	15.19746	2.5553	**13.7708**	*
T05	30	−36.8454	−32.6504	−35.0567	−34.9945	1.079044	**−36.7596**	***
T06	30	−29.1661	−21.2481	−27.4298	−26.9179	2.593203	−29.1637	≈
T07	20	0.5	0.71221	**0.53969**	0.55768	0.06638	1.06219	***
T08	7	220	220	220	220	0	220	≈
T09	126	1141.55	2921.4	1979.1	2011.64	440.807	1803.46	≈
T10	12	−21.8425	−11.1765	−13.8627	−15.5159	3.550713	−21.6445	***
T11.1	120	50666.7	53375.8	52112	52059.7	723.4086	51939.1	≈
T11.2	240	1069190	1076830	1072060	1072358	1750.415	1072290	≈
T11.3	6	15444.2	15476.5	15444.2	15452.59	13.46269	**15444.2**	**
T11.4	13	18022.2	18236.6	18108.9	18113.41	45.89656	18084.7	≈
T11.5	15	32692.4	32741.6	**32692.6**	32703.49	19.68271	32741.2	***
T11.6	40	122499	126802	123794	123845	884.3795	123953	≈
T11.7	140	1807750	1886540	**1852760**	1850503	17300.15	1859350	*
T11.8	96	929874	935719	931894	932289	1587.75	932304	≈
T11.9	96	936967	946713	940209	940333	2999.30	939335	≈
T11.10	96	930055	935881	932312	932357	1513.64	932648	≈
T12	26	9.26196	16.4275	**14.7501**	14.5522	1.55348	15.7757	**
T13	22	8.60885	18.3294	**14.2471**	13.4264	2.45308	14.8376	*

strategies in cSO is $K = 8$, the parameter to reset probabilities of strategies is $\delta = 1/(5 \times K) = 0.025$. The portion of the population to compute Eigenvectors is set to a recommended value $ps = 0.5$.

4   Results and Discussion

Basic characteristics from the results of the proposed cSO algorithm are in Table 1. On the right side, there are median values of the original jSO algorithm to make a simple comparison. In the last column, there are the significance values of the Wilcoxon rank-sum test denoted as follows: '***' ($p < 0.001$), '**' ($p < 0.01$), '*' ($p < 0.05$), and '≈' otherwise. The original jSO performs better in eight out of 22 real-world problems whereas the proposed cSO is better in ten out of 22 problems. Focusing only on significant differences, jSO performs significantly better in four and new cSO in six out of 22 problems. In the remaining 12 problems, both methods provide similar results.

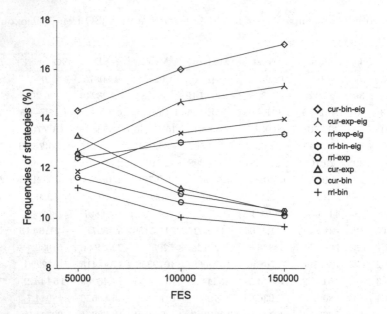

Fig. 1. Average frequencies of using strategies in cSO on all real-world problems.

High performance of cSO results in the research question 'Which part of the proposed method provides better results?'. For this purpose, the frequencies of using the eight strategies in the competition are analysed. These frequencies estimate of the use of strategies. Percentage frequencies of the use strategies are computed from results in three stages, $FES = 50000, 100000, 150000$ and average frequencies from all problems are illustrated in Fig. 1.

The distribution of the frequencies is changed during the search. Similar frequencies of the strategies at the early stage are gradually divided into two groups. Strategies employing Eigenvector coordinate system are used more frequently, and classic strategies are preferred rarely. Therefore the distribution of the frequencies of the use of strategies for each real-world problem is in Table 2. The first letter of strategies abbreviations is based on mutation variant ('c' for current-to-$p$best and 'r' for randrl/1). The second letter is given by employed crossover ('b' for binomial and 'e' for exponential). If strategy uses Eigenvector coordinate system, more 'e' is at the end of abbreviation. For problems with $D < 10$, mutation randrl/1 is preferred, problems with $10 < D \leq 40$, the original mutation with exponential crossover is used most often, and for problems where $40 \leq D$, the original strategy with Eigenvector transformation is preferred.

Table 2. Frequencies of four strategies without and with Eigenvector coordinate transformation used in cSO.

F	D	FES = 50000				FES = 100000				FES = 150000			
		cb	ce	rb	re	cb	ce	rb	re	cb	ce	rb	re
T01	6	4.9	8.3	16.4	**21.3**	5	8.3	17.1	**22.1**	5	8.3	17.2	**22.1**
T02	30	6.5	**33.9**	6.1	23.5	9.5	**27.5**	7.8	17	10.5	**21.2**	6.7	11
T03	1	5.4	5.1	20.1	**20.7**	5.4	5.1	20.1	**20.7**	5.3	5	20.2	20.8
T04	1	5.2	5.0	21.3	20.2	5.2	5.0	21.3	20.2	5.2	5.0	21.3	20.2
T05	30	8.8	**23.3**	10.6	18.1	7.6	13.2	9.6	9.8	7.3	11	7.7	7.7
T06	30	5.3	17.3	**17.4**	11.8	5.8	9.1	10.9	6.2	5.8	8	9.4	5.8
T07	20	7.8	19.5	10.5	9.8	7	12	8.3	6.8	6.6	10.6	7.6	6.1
T08	7	6.9	15.4	20.6	**26.9**	6.9	15.4	20.6	**26.9**	6.9	15.4	20.6	**26.9**
T09	126	25.5	9.9	9.5	5.3	24.1	10	6.9	4.8	21.8	9.4	6.1	4.6
T10	12	5.3	7.2	12.4	13.8	5.3	6.5	10.6	11.2	5.7	6.8	10.9	11.2
T11.1	120	23.2	12.4	3.2	5.1	20.8	14.5	3	6.4	17.7	13.6	3.1	6
T11.2	240	26.4	11.6	4	4.2	24.7	12.3	3.7	4.7	20.5	12.1	3.5	4.9
T11.3	6	6.8	6.9	11.4	13.8	6.4	6.5	11.5	14.1	6.3	6.4	11.5	14.2
T11.4	13	4	4	8.4	8.5	4.5	4.4	7.3	7.1	4.8	4.7	7.5	7
T11.5	15	9.8	11.9	12.2	10	8.2	9.8	10.3	8.7	8.1	9.5	10	8.4
T11.6	40	14.5	8.2	10.4	7.1	10.6	6.5	7.6	5.7	10.2	6.4	7.3	5.5
T11.7	140	21.4	9.2	9.1	5.1	18	8.1	7.8	4.6	17	7.9	7.9	4.8
T11.8	96	18.2	8.6	9.7	6.3	14.9	7.3	8	5.2	14.3	7.1	7.7	5.1
T11.9	96	19	9.4	8	5.8	15.9	8.3	7.2	5.3	14.8	7.9	7.1	5.3
T11.10	96	13.9	8.9	9.4	6.3	10.9	7.5	7.8	5.4	10.4	7.2	7.5	5.3
T12	26	6	**30.6**	6.7	19.3	6.8	**26.9**	6.1	18.3	7.5	**21.2**	5.8	14.7
T13	22	8.4	**24.1**	11.5	14.2	8.3	**20.4**	9.1	10.7	8	**18.8**	8.2	9.3
F	D	cbe	cee	rbe	ree	cbe	cee	rbe	ree	cbe	cee	rbe	ree
T01	6	6.2	6.2	18.9	17.8	5.9	5.9	18.3	17.3	5.9	5.9	18.3	17.3
T02	30	6.5	10.4	6	7.1	9.8	12.6	8.4	7.5	15.2	16	10.2	9.2
T03	1	5.8	6.1	18.4	18.3	5.7	6.1	18.6	18.2	5.6	6	18.8	18.4
T04	1	4.3	4.5	17.7	**21.8**	4.3	4.5	17.7	**21.8**	4.3	4.5	17.7	**21.8**
T05	30	8.7	14.4	8.7	7.5	12.1	**20.6**	13.3	13.9	15.8	**22.6**	13.9	14
T06	30	5.4	16.7	16.7	9.4	11.2	**22.7**	16.7	17.2	14.1	**22.5**	16.8	17.7
T07	20	8.6	**20.2**	11.9	11.8	13.1	**24.2**	13.4	15.3	14.6	**24.8**	13.3	16.5
T08	7	5.4	5	9.2	10.7	5.4	5	9.2	10.7	5.4	5	9.2	10.7
T09	126	**27.6**	7.9	9.1	5.2	**30.6**	9.2	8.6	5.7	**31.2**	10.8	8.9	7.2
T10	12	8.3	14.8	14.5	**23.8**	9.6	17	14.5	**25.2**	9.8	16.8	14.1	**24.6**
T11.1	120	**24.5**	18.7	5.2	7.7	**23.9**	18.2	5.3	7.9	**24.9**	19	6.7	9.1
T11.2	240	**28.9**	13.6	5.1	6.2	**29.6**	13.6	5.5	5.9	**29.5**	15.2	6.8	7.4
T11.3	6	9.3	10.7	17.5	**23.7**	8.9	10.3	17.9	**24.4**	8.8	10.2	18	**24.5**
T11.4	13	11.4	20.5	19.1	**24.2**	14.7	20.3	18.5	**23.1**	15.2	19.8	18.4	**22.5**
T11.5	15	13.1	**15.5**	12.3	15.2	14.1	**18.3**	12.9	17.6	14.5	**18.5**	13	18
T11.6	40	**18.6**	17	13.6	10.6	21	**21.7**	14.3	12.5	21.8	**21.8**	14.4	12.6
T11.7	140	**25.9**	11.5	10.8	7	**25.7**	15.8	10.4	9.5	25	16.5	10.5	10.5
T11.8	96	**26.1**	11.1	13.8	6.1	**29.1**	13.7	14.5	7.4	**29.5**	14.1	14.5	7.8
T11.9	96	**24.9**	8.7	16.3	8	**25.2**	11.4	16.4	10.2	**24.7**	12.2	16.6	11.3
T11.10	96	**27.8**	12.9	14.1	6.7	**28.8**	15.8	15	8.8	**29.4**	16.1	14.9	9.1
T12	26	6.8	16.2	5.6	8.6	9	16.6	7	9.3	13.4	18.3	8.3	10.8
T13	22	8.3	13.2	10.2	10	11.4	16.4	12	11.6	13.1	17.8	12.5	12.3

5   Conclusion

The newly proposed variant of the efficient jSO algorithm employing competition of eight strategies (cSO) provides very good performance as it significantly outperforms the original method in six out of 22 real-world problems, and loses in four problems. This information is important because the original jSO belongs to the best performing state-of-the-art methods, and the proposed mechanism enables us to improve it. The analysis of the employed strategies provides clear information that the original strategy is never used most frequently. For low-dimensional problems, strategies with the newly used randrl/1 mutation are the most successful. In the case of middle dimensionality, the newly used exponential crossover is preferred, and for high dimension levels, the Eigenvector coordinate system provides the best results. Comparing cSO with the winner of CEC 2011 competition (GA-MPC [9]), cSO performs better in ten problems and GA-MPC in eight problems. In further research, a more adaptive mechanism of use and elimination of the employed strategies during the search will be studied.

References

1. Brest, J., Maučec, M.S., Bošković, B.: Single objective real-parameter optimization: algorithm jSO. In: 2017 IEEE Congress on Evolutionary Computation (CEC), pp. 1311–1318 (2017)
2. Bujok, P., Poláková, R.: Migration model of jSO algorithm. In: 2018 25th International Conference on Systems Signals and Image Processing (IWSSIP). IEEE, New York (2018). IEEE Slovenia Section, University of Maribor
3. Bujok, P., Poláková, R.: Eigenvector crossover in the efficient jSO algorithm. MENDEL Soft Comput. J. **25**, 65–72 (2019)
4. Bujok, P.: Tvrdík: enhanced success-history based parameter adaptation for differential evolution and real-world optimization problems. In: Papa, G., Mernik, M. (eds.) Bioinspired Optimization Methods and Their Applications, BIOMA, Bled, Slovenian, pp. 159–171 (2016)
5. Bujok, P., Tvrdík, J.: A comparison of various strategies in differential evolution. In: Matoušek, R. (ed.) MENDEL: 17th International Conference on Soft Computing, Brno, Czech Republic, pp. 48–55 (2011)
6. Bujok, P., Tvrdík, J., Poláková, R.: Evaluating the performance of SHADE with competing strategies on CEC 2014 single-parameter test suite. In: 2016 IEEE Congress on Evolutionary Computation (CEC), pp. 5002–5009 (2016)
7. Das, S., Mullick, S.S., Suganthan, P.N.: Recent advances in differential evolution-an updated survey. Swarm Evol. Comput. **27**, 1–30 (2016)
8. Das, S., Suganthan, P.N.: Problem definitions and evaluation criteria for CEC 2011 competition on testing evolutionary algorithms on real world optimization problems. Technical report, Jadavpur University, India and Nanyang Technological University, Singapore (2010)
9. Elsayed, S.M., Sarker, R.A., Essam, D.L.: GA with a new multi-parent crossover for solving IEEE-CEC2011 competition problems. In: 2011 IEEE Congress of Evolutionary Computation (CEC), pp. 1034–1040 (2011)
10. Kotyrba, M., Volna, E., Bujok, P.: Unconventional modelling of complex system via cellular automata and differential evolution. Swarm Evol. Comput. **25**, 52–62 (2015)

11. Storn, R., Price, K.V.: Differential evolution - a simple and efficient heuristic for global optimization over continuous spaces. J. Glob. Optim. **11**, 341–359 (1997)
12. Tanabe, R., Fukunaga, A.S.: Improving the search performance of SHADE using linear population size reduction. In: 2014 IEEE Congress on Evolutionary Computation (CEC), pp. 1658–1665 (2014)
13. Tvrdík, J.: Competitive differential evolution. In: Matoušek, R., Ošmera, P. (eds.) MENDEL 2006, 12th International Conference on Soft Computing, pp. 7–12. University of Technology, Brno (2006)
14. Wang, Y., Li, H.X., Huang, T., Li, L.: Differential evolution based on covariance matrix learning and bimodal distribution parameter setting. Appl. Soft Comput. **18**, 232–247 (2014)
15. Wu, G., Mallipeddi, R., Suganthan, P.N.: Ensemble strategies for population-based optimization algorithms - a survey. Swarm Evol. Comput. **44**, 695–711 (2019)
16. Zamuda, A., Sosa, J.D.H.: Success history applied to expert system for underwater glider path planning using differential evolution. Expert Syst. Appl. **119**, 155–170 (2019)

Neural Swarm Virus

Thanh Cong Truong[1]([✉])[iD], Ivan Zelinka[1][iD], and Roman Senkerik[2][iD]

[1] Faculty of Electrical Engineering and Computer Science,
VSB-Technical University of Ostrava,
17. listopadu 2172/15, 708 00 Ostrava-Poruba, Czech Republic
{cong.thanh.truong.st,ivan.zelinka}@vsb.cz
[2] Faculty of Applied Informatics, Tomas Bata University in Zlin,
T. G. Masaryka 5555, 760 01 Zlin, Czech Republic
senkerik@utb.cz

Abstract. The dramatic improvements in computational intelligence techniques over recent years have influenced many domains. Hence, it is reasonable to expect that virus writers will taking advantage of these techniques to defeat existing security solution. In this article, we outline a possible dynamic swarm smart malware, its structure, and functionality as a background for the forthcoming anti-malware solution. We propose how to record and visualize the behavior of the virus when it propagates through the file system. Neural swarm virus prototype, designed here, simulates the swarm system behavior and integrates the neural network to operate more efficiently. The virus's behavioral information is stored and displayed as a complex network to reflect the communication and behavior of the swarm. In this complex network, every vertex is then individual virus instances. Additionally, the virus instances can use certain properties associated with the network structure to discovering target and executing a payload on the right object.

Keywords: Swarm virus · Swarm intelligence · Neural network · Malware · Computer virus · Security

1 Introduction

Malicious software or malware is a software or program that infiltrates or damage a computer system without consent and without informing the system owner. Technically, researchers used this term to express a variety of forms of malicious programs that including virus, worm, trojan horse, exploits, botnet, retrovirus [33]. Nevertheless, the most common type of malware is computer virus, a terminology given by Cohen in [9]. While the initial start of malware was just unharmful software that causing mild annoyance, later it was created with the financial goal or even a virtual weapon for cyberwar as discuss in [22]. Today malware creation has become a commercial industry with revenues to billion dollars.

A. Zamuda et al. (Eds.): SEMCCO 2019/FANCCO 2019, CCIS 1092, pp. 122–134, 2020.
https://doi.org/10.1007/978-3-030-37838-7_12

The combat between malware and anti-malware is an endless war in which cyber-threat actors adopt new techniques to thwart detection while the cyber-defenders try to find effective measures to prevent this. The anti-malware community continues to adopt various kind of heuristics (such as artificial neural networks [28]) to recognize new and unknown malicious codes. In contrast, malware authors continuously attempt to find different methods to surpass the defense system.

A considerable amount of literature has been published on discussing the dynamics and spreading of malware. The primary trend in this research area is the dynamics of malware as well as the malware behaviors in real-world networks. As an example, the authors in [26] investigated the computer virus infection by adapting the epidemiologically compartmental models. They have drawn a mathematical model and identified potential edges where contagion could occur. Meanwhile, the authors in [40] proposed a novel virus heterogenous propagation model and its optimal control problem in which they considered the joint impact of countermeasure and network topology on virus diffusion and optimal dynamic countermeasure. Simultaneously, researchers in [31] presented a moderate epidemiological model based on the fractional epidemiological model to describe computer viruses with an arbitrary order derivative having a non-singular kernel.

Malware behaviors are also investigated by scientists in [23, 27, 30]. In [23], the authors investigated how the infection rates affect virus propagation by combining the Susceptible Infected (SI) and the Susceptible Infected Recovered (SIR) model then adopted this to the Barabasi–Albert network. Meanwhile, Parsaei et al. [27] combined Lyapunov functions with the Volterra-Lyapunov matrix properties to apply for a computer virus propagation model. In the meantime, scientists in [30] adopted the Routh-Hurwitz criterion and Lyapunov functional approach on a computer virus propagation model based on the kill signals called SEIR-KS.

In other research, Noreen et al. [25], Meng et al. [24] propose the framework for malware evolution base on the evolutionary computation to evolve computer and Android malware, respectively. Meanwhile, the authors in [8] proposed to exploit an Evolutionary Algorithm (EA) to auto-generate malware. Other research by Kudo et al. [20, 21] introduced the botnets that adopting machine learning (ML) methods to predict vulnerabilities and evolve itself autonomously.

Recently, the fusion between malware and computational intelligence is another research trend that concern by scientists. For example, Geigel in [15, 16] proposed to apply neural network with supervised and unsupervised learning to encode the trojan. Meanwhile, several recent studies [1–3, 17, 18, 35] researched how to evade the anti-malware with ML engines. In a later study, Zelinka et al. [38] design a swarm virus prototype, which mimics a swarm system behavior.

The objective of this paper is to outline a possible dynamics swarm malware adopting neural network techniques as well as its structure, and functionality. Our research's aim is to develop a prototype of neural swarm malware instead of fully functional one, this prototype is lacking payload, and the contagion can be controlled.

The remains of this article proceed as follows. Section 2 presents essential background information. In Sect. 3, we present the neural swarm malware, its structure, and functionality. The following sections report the setup and results of the experiment. We conclude our paper in Sect. 6.

2   Background: Malware, Swarm-Based Intelligence, and Neural Network

2.1   Malware

Malware is a general term for many types of malicious software, including virus, worm, trojan horse, rootkit, spyware, ransomware, and others. In particular, the most pervasive form of malware is the computer virus. The term "computer virus" is derived from and is in some sense similar to a biological virus [32]. It is an automated program capability to self-replicate and attack various host files. The host files can be executable files, boot code, device drivers, or files that unable to execute directly but through specific applications (Microsoft Word, Visual Basic scripts, and others). Once the infected hosts run, it also executes the virus code, and the virus propagates further by self-reproduce and attached to another host.

The idea of computer virus can be traced back to John von Neumann in the 1950s with cellular automata and self-replicating programs [14], yet the first working computer virus appeared much later. The first recorded instance of a self-replicating computer program is the Creeper, developed by Bob Thomas in 1971 [8]. Later, in 1983, Fred Cohen demonstrated a program that capability infected a computer, replicated, and spread to other computers which lead to the born of the term "computer virus" [9]. In 1986, the first PC virus named Brain was observed in the wild [33], it was widely considered the first real malware.

In addition to the virus, other terminologies are used to describe specific malware, indicating their specific purpose, spreading strategy or behavior. For instance, a worm is a virus that can spread over the network by exploiting vulnerabilities. Trojan horses disguise as legitimate programs and executed malicious routines or files on the host. Spywares are software used to hijacks personal and confidential information. Adware is generic advertising-supported software automatically delivers advertisements in the system. Ransomware is designed to lock a computer system or crypt the victim's data to extort money from the victim. Rootkits take control of the infected machine by gaining the highest privileges possible on the machine. Botnets are a pack of malware remotely controlled from one server.

2.2   Swarm Intelligence

Swarm intelligence is a computational technique in which many individuals coordinate using decentralized control to solve the problem. It focuses on the collective behavior exhibited by the interaction of the individuals. Swarm intelligence provides another way to design "intelligent" systems. [6]. Regarding advancing in swarm-based intelligence and technologies, it is logical to expect that, shortly swarm intelligence will be utilized in used as both attack and defenses tools in cyberspace. Very promising algorithms (regarding malware could be derivative from) are algorithms that inspired by nature such as Genetic Algorithm [34]. Another interesting algorithm (regarding command and control worms like Botnets are) are swarm algorithms such as ant colony optimization (ACO) [11] followed by Particle Swarm Optimization (PSO) [12,19] and Self Organizing Migration Algorithm (SOMA) [10,36,39].

2.3   Neural Network

Artificial neural network (ANN), mainly known as Neural Network (NN) is a computational model inspired by the structure and operation of the biological neural network. The core of this paradigm is to connect the neurons (or nodes) into a computer network that can perform complicated tasks. Each node connected to another node through a connection and each link has a "weight" with the information of the input signal. The weight represents the importance (or strength) level of the input data and is used by the network to solve the problem. Also, each neuron has a state, which indicates that the activation capability of that neuron. These neurons can generate an output by combining the input signals and activation rule.

The general architecture of an ANN consists of three components: the input layer, hidden layer, and output layer. In this architecture, the hidden layer consists of neurons that receive input data from the neurons in the previous layer and convert these inputs for subsequent processing layers. Moreover, there may be many hidden layers in an ANN.

ANN learning paradigms can be categorized by the method of training carried out as supervised, unsupervised, and reinforcement learning. Supervised learning is a process to learn the mapping function from the input to the output. Unsupervised learning model identifies the pattern class information with only input data and no corresponding output variables. Reinforcement learning learns by using the reward and penalty system. It learns through an interactive environment by trial and error using feedback from its actions.

3   Neural Network Swarm Virus Creation

3.1   General Idea

Malware has advanced dramatically since the first appeared of a computer virus. Malware authors adopted various techniques such as encryption, oligomorphism, polymorphism, metamorphism, obfuscation [29], armouring (armoured viruses) [13] to evade anti-malware tools. Furthermore, the latest virus can be controlled via command and control (C&C) infrastructure as in the case of [22] virus. The weak point of this structure is that it can be immobile if the control center was destroyed.

Our approach is stem from the idea of eliminating C&C center in the botnet structure. Thus, we propose to fuse swarm base intelligence, neural network, and a classical computer virus to form a neural swarm virus. This virus can mimic the behavior of the biological swarm systems, which usually do not have a dominant central communication point. Our research aims to develop a prototype of a neural network swarm virus without payload while the contagion is controllable and limited. Generally speaking, our research is differential with contemporary research papers cited above, which based on mathematical models and numerical simulations. In contrast, our paper is based on real experimentations with prototype malware in a secured virtual environment, real-time observations, data recording, visualizations, and some fundamental analysis.

Our idea is to present a combination of swarm base intelligence, neural network, and a traditional virus to develop a new kind of virus. Technically, this virus consists of instances (individuals of the population) that form a swarm (population) that propagate in the computer file system. The individuals in swarm communicate via the command line (when shifting from file to file) and amongst themselves. Then, a network is created basing on the exchange information mechanism and swarm motion through system file. This network represents their physical presence on different hosts.

3.2 Spreading Mechanisms

The spreading of malware is quite complicated and mainly depend on the kind of malware (e.g., virus, trojan, worm) and the environment (inside the computer or network). The distribution of malicious programs has expanded beyond traditional ways such as from removable media, downloads files from the Internet, or e-mail attachments to more sophisticated approaches like drive-by downloads from a compromised website or using social engineering.

In fact, the malware could use various infection techniques to move inside the PC environment such as prepending, appending, or inserting it into an executable file. In our virus prototype, we adopt the prepending technology. With this technology, the virus attaches itself to the start of the host so that it will execute first when the program starts. Furthermore, the spreading strategy resembles the classical worm called "Rabbit" [4]. In this strategy, the virus will erase its copy on previously visited files when moving to a new one. Consequently, the host will be recovered to the original state. In this view, the virus's behavior is similar to evolutionary or swarm algorithms, whose individuals jump over the search space. This strategy controls the spread of the virus to avoid excessive population growth, which may cause system slowdown and lead to detection. Hence, the population of the swarm virus will remain unchanged during the contagion.

3.3   Virus Structure and Functionality

The NN swarm virus is a self-replicating structure, consisting of components to do its task. In our experiments reported in this paper, the virus components are in Fig. 1

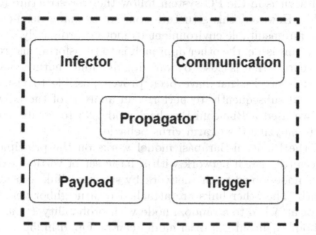

Fig. 1. The proposed virus structure

The functionality of each component is as follows:

- Infector: This component is responsible for some tasks as follow. The first task is copying the virus and attaching it to a suitable host. The second is healing the host file after moving to a new location. The next one is to check whether the file is infected or not; if the file is infected, then there no need to infect it again.
- Propagator: This component is the most important because it decides how the virus is propagating. It will indicate where to move (i.e., which file to infect) by evaluating the file. The evaluating is based on neural network methods to find fitness. In our experiments, the fitness value is calculated by the size of the file. However, other features of the file could be used, for example, most recently modified file, type of files, or the content of the file.
- Communication: Communication is responsible for communicating in the swarm. The virus-instance interact with each other to decide which one should be activated through the command line arguments. What is more, it allows the virus updating the storing information such as locations of all virus instances, the network topology when moving to a new file. That means, the entire communication traffic going from virus instance to another one without the central communication point.
- Payload: In this prototype, the payload is to test the swarm functionality. We do not implement any destructive payload.
- Trigger: The trigger is to launch the payload at a given condition. The trigger can be set to act on a given condition.

3.4   Virus Behavior Patterns

The most crucial factor in observing and analyzing swarm virus is behavior patterns. Behavior patterns are accessible data and are often part of a large data set that records malware behavior in the system. Technically, the behavior – movement of a virus in the PC system follow the tree structure (i.e., moving from file to file). However, this structure consists of many dead-ends and no-cycles, which is an unsuitable environment to move coordinately.

To overcome this issue, the other approach is to transform this tree structure into another structure that is close to swarming dynamics nature. Such networks can be complex networks that have been proven possible to visualize swarm dynamics [37] and subsequently to perform an analysis of the network. Thus, in this paper, we used a Bianconi-Barabási model [5] to create a network for visualizing and analyzing the swarm virus behavior.

In general, the Bianconi-Barabási model works on the principle of how to add new vertices to a small network with a basic set of vertices exists. In this model, the new node joined to the network by setting a link to a random node j of the network. The other links are attached to a neighbor vertex of j with probability p, or attached to a random node with probability $1 - p$. More detail information about Bianconi-Barabási model can be found in [5].

3.5   Adopting Neural Network to Target Specific Object

Scientists have foreseen the misuse of Artificial intelligence (AI) in the report [7], which mentioned that AI-based approaches could be used in malware. Inspired by this idea, we adopt an AI technique in our swarm virus. In our prototype, the use of a neural network will make the "trigger conditions" to infect a file or not.

The virus is trained to search in the system until the intended target is reached. Later, the neural network generates the signal needed to perform the infection (moving). In our experiments, we use the file size to identify the target. However, other attributes could also be used, for example, system-level features. This method will make it difficult for malware analysts to reverse the neural network reverse engineer the neural network to discover the specifics of the target.

Generally speaking, the swarm virus leverages the black-box nature of the NN model to camouflage the trigger condition. The NN model in the malware will replace the traditional "if-then" trigger condition so that it is tough to decipher. Technically, this method will be an extreme challenge for malware analysts to figure out what category of the target that malware is looking for, or what is the specific target to trigger condition. Figure 2 illustrated the comparison of traditional and NN-powered targeted attack.

4   Experiment Setups

In order to test our hypothesis, we developed a prototype of the virus in a high programming language - C#, which is very convenient laboratory experiments

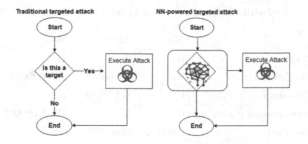

Fig. 2. Comparison of traditional and NN-powered targeted attack

and research. In our experiments, we tested the virus under Windows 10 environment in the virtual machine.

The malware behavior pattern is a significant feature when studying malicious software. To record and visualize the behavior of this swarm virus, we used a slightly adjusted Bianconi-Barabási model for creating scale-free networks. In Bianconi-Barabási model, two parameters need to be set up is the probability p and the number of links to attach for each new vertex m. In our experiments, the parameter p was set at least to 0.5. The probability 0.5 means that the ratio of connecting to neighbors or random vertex is equal. Another parameter to set was the m parameter, which represents the number of connections it will initially have. Experiments showed that the value higher than four would give better performance. In the network, each edge weights $1/d$, where d is the distance of the file, measured by the number of folders between files (vertices) + 1. The complete set of recommended values are shown in Table 1.

Table 1. Recommended parameters of NN swarm virus.

Parameter	Value
Number of virus instance	5–up to user
Visited host length	20–up to user
p	0.4–0.9
m	4–10

5   Results

We tested our virus prototype in Windows 10 environment on a single PC with 16 Gb RAM, Intel Core i7 8th generation. The experiment was repeated 100 times. In each experiment, the virus activities were recorded, including the file name, path, and its behaviors (such as write a file, delete a file, movement). The virus behavior in the system had been analyzed for networks attributes like degree, closeness centrality, betweenness centrality, and page rank. The results are visualized in Figs. 3, 4, 5, and in Table 2.

Table 2. Swarm virus network centralities

	Min	Median	Max
Degree	0	1	47
Weighted degree	0	31.666667	258.909091
Closeness centrality	0	0.177778	1
Betweenness centrality	0	222	1772.188095
Pagerank	0.003024	0.003651	0.069334

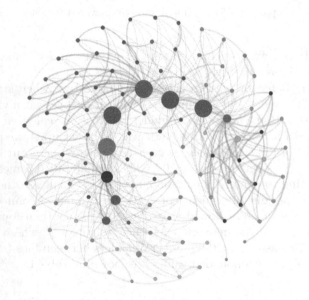

Fig. 3. Network example 1. The size of each node represents an importance of a file

In our experiments, we created a swarm consist of 5 individuals, so that means we had five virus-instance. In each experiment, an individual jumped from file to file (infected a file) 20 times in total. Although the population of the swarm was small and the movement was low, we recorded and visualized a fascinating behavior of swarm. Figures 3, 4 and 5, illustrated the behavior of the swarm. In these figures, the node size and color presented the file importance such as file size, or the frequency of file infections by a virus. The edges represent, which file was visited by virus instance.

The degree of a node is the number of relation (edge) it has, which is the sum of edges for a node. Technically, a node with a higher degree has more neighbor than the others. In Figs. 3, 4 and 5 some nodes were bigger than other nodes, which mean they are more important than the rest.

Similarly, the more significant node has a higher PageRank. As a result, if a node has a higher PageRank, then it has more probability of visiting. The goal is to visit the most critical nodes in the network.

Fig. 4. Network example 2

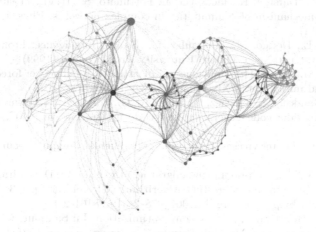

Fig. 5. Network example 3

6 Conclusion

In this paper, we presented the idea of "hypothetical" computer swarm virus, which adopting neural network to enhance its power. This research has discussed the virus structure, functionality, and capabilities. Furthermore, we have proposed a method to visualize, measure, and analyze the behavior of the swarm-based virus in the form of a complex network. The results of this study shall be useful for the understanding of the behavior of new generation malware as well as future computer technology protection.

Acknowledgement. The following grants are acknowledged for the financial support provided for this research: Grant of SGS No. SP2019/137, VSB Technical University of Ostrava. This work was also supported by the Ministry of Education, Youth and Sports of the Czech Republic within the National Sustainability Programme Project no. LO1303 (MSMT-7778/2014), further by the European Regional Development Fund under the Project CEBIA-Tech no. CZ.1.05/2.1.00/03.0089.

References

1. Anderson, H.S., Kharkar, A., Filar, B., Evans, D., Roth, P.: Learning to evade static PE machine learning malware models via reinforcement learning. arXiv:1801.08917 (2018)
2. Anderson, H.S., Kharkar, A., Filar, B., Roth, P.: Evading machine learning malware detection. Black Hat (2017)
3. Anderson, H.S., Woodbridge, J., Filar, B.: DeepDGA: adversarially-tuned domain generation and detection. In: Proceedings of the 2016 ACM Workshop on Artificial Intelligence and Security, pp. 13–21. ACM (2016)
4. Aycock, J.: Computer Viruses and Malware, vol. 22. Springer, Heidelberg (2006). https://doi.org/10.1007/0-387-34188-9
5. Bianconi, G., Darst, R.K., Iacovacci, J., Fortunato, S.: Triadic closure as a basic generating mechanism of communities in complex networks. Phys. Rev. E **90**(4), 042806 (2014)
6. Bonabeau, E., Dorigo, M., Theraulaz, G.: Swarm Intelligence: From Natural to Artificial Systems. No. 1. Oxford University Press, Oxford (1999)
7. Brundage, M., et al.: The malicious use of artificial intelligence: forecasting, prevention, and mitigation. arXiv preprint arXiv:1802.07228 (2018)
8. Cani, A., Gaudesi, M., Sanchez, E., Squillero, G., Tonda, A.P.: Towards automated malware creation: code generation and code integration. In: SAC, pp. 157–160 (2014)
9. Cohen, F.: Computer viruses: theory and experiments. Comput. Secur. **6**(1), 22–35 (1987)
10. Zelinka, I.: SOMA—self-organizing migrating algorithm. In: Davendra, D., Zelinka, I. (eds.) Self-Organizing Migrating Algorithm. SCI, vol. 626, pp. 3–49. Springer, Cham (2016). https://doi.org/10.1007/978-3-319-28161-2_1
11. Dorigo, M., Birattari, M.: Ant colony optimization. In: Sammut, C., Webb, G.I. (eds.) Encyclopedia of Machine Learning. Springer, Boston (2011). https://doi.org/10.1007/978-0-387-30164-8
12. Eberhart, R., Kennedy, J.: A new optimizer using particle swarm theory. In: Proceedings of the Sixth International Symposium on Micro Machine and Human Science, MHS 1995, pp. 39–43. IEEE (1995)
13. Filiol, E.: Strong cryptography armoured computer viruses forbidding code analysis: the Bradley virus. Ph.D. thesis, INRIA (2004)
14. Filiol, E.: Computer Viruses: From Theory to Applications. Springer, Heidelberg (2006). https://doi.org/10.1007/2-287-28099-5
15. Geigel, A.: Neural network Trojan. J. Comput. Secur. **21**(2), 191–232 (2013)
16. Geigel, A.: Unsupervised learning Trojan. Ph.D. thesis, Nova Southeastern University (2014)
17. Grosse, K., Papernot, N., Manoharan, P., Backes, M., McDaniel, P.: Adversarial perturbations against deep neural networks for malware classification. arXiv preprint arXiv:1606.04435 (2016)
18. Hu, W., Tan, Y.: Generating adversarial malware examples for black-box attacks based on GAN. arXiv preprint arXiv:1702.05983 (2017)

19. Kennedy, J.: Swarm intelligence. In: Zomaya, A.Y. (ed.) Handbook of Nature-Inspired and Innovative Computing, pp. 187–219. Springer, Boston (2006). https://doi.org/10.1007/0-387-27705-6_6
20. Kudo, T., Kimura, T., Inoue, Y., Aman, H., Hirata, K.: Behavior analysis of self-evolving botnets. In: 2016 International Conference on Computer, Information and Telecommunication Systems (CITS), pp. 1–5. IEEE (2016)
21. Kudo, T., Kimura, T., Inoue, Y., Aman, H., Hirata, K.: Stochastic modeling of self-evolving botnets with vulnerability discovery. Comput. Commun. **124**, 101–110 (2018)
22. Kushner, D.: The real story of stuxnet. IEEE Spectr. **3**(50), 48–53 (2013)
23. Lazfi, S., Lamzabi, S., Rachadi, A., Ez-Zahraouy, H.: The impact of neighboring infection on the computer virus spread in packets on scale-free networks. Int. J. Mod. Phys. B **31**(30), 1750228 (2017)
24. Meng, G., et al.: Mystique: evolving android malware for auditing anti-malware tools. In: Proceedings of the 11th ACM on Asia Conference on Computer and Communications Security, pp. 365–376. ACM (2016)
25. Noreen, S., Murtaza, S., Shafiq, M.Z., Farooq, M.: Evolvable malware. In: Proceedings of the 11th Annual Conference on Genetic and Evolutionary Computation, pp. 1569–1576. ACM (2009)
26. Pan, W., Jin, Z.: Edge-based modeling of computer virus contagion on a tripartite graph. Appl. Math. Comput. **320**, 282–291 (2018)
27. Parsaei, M.R., Javidan, R., Kargar, N.S., Nik, H.S.: On the global stability of an epidemic model of computer viruses. Theory Biosci. **136**(3–4), 169–178 (2017)
28. Prasse, P., Machlica, L., Pevný, T., Havelka, J., Scheffer, T.: Malware detection by analysing encrypted network traffic with neural networks. In: Ceci, M., Hollmén, J., Todorovski, L., Vens, C., Džeroski, S. (eds.) ECML PKDD 2017. LNCS (LNAI), vol. 10535, pp. 73–88. Springer, Cham (2017). https://doi.org/10.1007/978-3-319-71246-8_5
29. Rad, B.B., Masrom, M., Ibrahim, S.: Camouflage in malware: from encryption to metamorphism. Int. J. Comput. Sci. Netw. Secur. **12**(8), 74–83 (2012)
30. Ren, J., Xu, Y.: A compartmental model for computer virus propagation with kill signals. Phys. A **486**, 446–454 (2017)
31. Singh, J., Kumar, D., Hammouch, Z., Atangana, A.: A fractional epidemiological model for computer viruses pertaining to a new fractional derivative. Appl. Math. Comput. **316**, 504–515 (2018)
32. Spafford, E.H.: Computer viruses as artificial life. Artif. Life **1**(3), 249–265 (1994)
33. Szor, P.: The Art of Computer Virus Research and Defense. Pearson Education, London (2005)
34. Whitley, D.: A genetic algorithm tutorial. Stat. Comput. **4**(2), 65–85 (1994)
35. Xu, W., Qi, Y., Evans, D.: Automatically evading classifiers. In: Proceedings of the 2016 Network and Distributed Systems Symposium, pp. 21–24 (2016)
36. Zelinka, I.: SOMA - self organizing migrating algorithm. In: Onwubolu, G.C., Babu, B.V. (eds.) New Optimization Techniques in Engineering. STUDFUZZ, vol. 141, pp. 167–217. Springer, Heidelberg (2004). https://doi.org/10.1007/978-3-540-39930-8_7
37. Zelinka, I., Chen, G.: Evolutionary Algorithms, Swarm Dynamics and Complex Networks: Methodology, Perspectives and Implementation, vol. 26. Springer, Heidelberg (2017). https://doi.org/10.1007/978-3-662-55663-4
38. Zelinka, I., Das, S., Sikora, L., Šenkeřík, R.: Swarm virus - next-generation virus and antivirus paradigm? Swarm Evol. Comput. **43**, 207–224 (2018)

39. Zelinka, I., Jouni, L.: Soma - self-organizing migrating algorithm. In: Mendel 2000, 6th International Conference on Soft Computing, Brno, Czech Republic, pp. 177–187 (2000)
40. Zhang, X., Gan, C.: Global attractivity and optimal dynamic countermeasure of a virus propagation model in complex networks. Phys. A **490**, 1004–1018 (2018)

Wrapper-Based Feature Selection Using Self-adaptive Differential Evolution

Dušan Fister[1]($\boxtimes$)(iD), Iztok Fister[2](iD), Timotej Jagrič[1], Iztok Fister Jr.[2](iD), and Janez Brest[2](iD)

[1] Faculty of Economics and Business, University of Maribor, Razlagova 14, 2000 Maribor, Slovenia
{dusan.fister1,timotej.jagric}@um.si
[2] Faculty of Electrical Engineering and Computer Science, University of Maribor, Koroška cesta 46, 2000 Maribor, Slovenia
{iztok.fister,iztok.fister1,janez.brest}@um.si

Abstract. Knowledge discovery in databases is a comprehensive procedure which enables researchers to explore knowledge and information from raw sample data usefully. Some problems may arise during this procedure, for example the Curse of Dimensionality, where the reduction of database is desired to avoid feature redundancy or irrelevancy. In this paper, we propose a wrapper-based feature selection algorithm, consisting of an artificial neural network and self-adaptive differential evolution optimization algorithm. We test performance of the feature selection algorithm on a case study of bank marketing and show that this feature selection algorithm reduces the size of the database and simultaneously improves prediction performance on the observed problem.

Keywords: Data preprocessing · Feature selection · Self-adaptive differential evolution jDE · NiaPy

1 Introduction

In the era of big data, where more and more data are analyzed, a Data Mining (DM) paradigm [40] has been proposed to stimulate the design and analysis of diverse DM methods to deal with sample data. A wide range of DM methods exist nowadays, from traditional regression analysis to Machine Learning (ML) and symbolic methods. Those can solve problems like classification, regression, clustering, association rule mining and others, in a supervised, unsupervised or reinforcement learning way.

The data are becoming more and more complex. The rise of the Industry 4.0 [19] and the rise of the mobile devices contribute inherently to such trend. As a result, the analysis of data is, according to the Curse of Dimensionality [5] becoming tougher. Typically, the range and volatility of values increases, generalization diminishes and outliers with missing data emerge. In order to avoid these issues, a data preprocessing step is necessary, which executes the data

© Springer Nature Switzerland AG 2020
A. Zamuda et al. (Eds.): SEMCCO 2019/FANCCO 2019, CCIS 1092, pp. 135–154, 2020.
https://doi.org/10.1007/978-3-030-37838-7_13

adjustment procedures to obtain improved predictive abilities of DM methods. Proper data preprocessing step typically tries to decrease or eliminate feature redundancy/irrelevancy and may thus bring lower costs and higher modeling efficiencies. In the following paper, we would like to empirically investigate the practical worth of the data preprocessing step, i.e. what are its benefits, and list its shortcomings.

Literature conceptualizes both the DM and data preprocessing as sub-parts of a more comprehensive Knowledge Discovery in Databases (KDD) procedure [15]. The latter formalizes six phases as shown below:

- problem specification, where the actual problem and the estimated outcome of the KDD are first addressed,
- problem understanding, where the problem tries to become explainable,
- data preprocessing, where the data are prepared, and the complexity is reduced by removing redundant or irrelevant items,
- data mining, where the actual model and task are determined to mine the preprocessed data and to extract discovered knowledge,
- evaluation, where the obtained results are interpreted and
- result exploitation, where the knowledge discovered is visualized for a report to form inference.

The purpose of the mentioned six phases of KDD is to extract hidden patterns, relations and interconnections among data. Data usually consist of explanatory variables (features), i.e., inputs, and the response variables (also target), i.e., outputs. Relations can then be studied for various inference and analysis applications, such as student performance [25], cost reduction [35], satisfying customer expectations [10], healthcare [4], forecasting [7] and others.

The more the interconnections exist in data, the more the redundancies arise. This affects the prediction performance negatively. Feature selection (FS) is a suitable process of eliminating some of the variables, to diminish or avoid redundancy [41]. Three FS methods exist for the supervised tasks [1,15]: filter-based method, wrapper-based method and embedded-based method. Filter-based method performs the FS separately from the learning algorithm [42], while wrappers do not. They use the learning algorithm to determine the quality of the selected subset [17]. Embedded methods, on the other hand, combine characteristics of the previous two [38]. The following survey outlines many FS applications [8].

Osanaiye et al. [27] show the application of filter-based method for feature elimination in the field of cloud computing. Authors state that number of features can be reduced efficiently from 41 to 13, by using their method. Apolloni et al. [3] claim that their FS methods decrease the number of relevant features for more than 99% on the microarray problem, on the basis of six datasets. Labani et al. [18] design a novel multivariate filter-based method FS for text classification problems, and prove that it can overcome other univariate and multivariate methods.

Mafarja and Mirjalili [24] outline a wrapper-based method using the whale optimization. They use the mutation and crossover operators to increase population diversity and tournament selection to select best individuals. Authors test their FS method on the 18 different datasets. For most of the datasets, they show that their method improves other filter-based FS methods. Al-Tashi et al. [2] deal with combining the hybrid grey wolf optimization (GWO) and particle swarm optimization (PSO) for FS problem. Ramjee and Gamal [31] experiment with wrapper-based FS, where they employ the relevance and redundancy scores. Based on their solution, they report the increase of accuracy from initial 93% to final 99% for the RadioML2016.10b dataset. Lu [23] agrees that data heterogeneity leads to spurious classification and thus proposes the embedded FS method as a remediation. Author uses the synthetic data and three benchmark datasets to confirm that his solution bears fruit, compared to traditional embedded-, or filter-based methods. Liu et al. [21] deal with the embedded FS on the imbalanced data, i.e. fraud detection and cancer diagnosis, where they propose a weighted Gini index GI-FS, specifically designed to handle imbalanced data. ECoFFeS [22] is a comprehensive user-friendly and standalone software intended for automated FS in drug discovery, since it incorporates a set of single-objective and multi-objective evolutionary computation algorithms. Wang et al. [39] shows the application of PSO algorithm for descriptor selection.

This paper is an extension of the [14], where we employed an FS procedure using the logistic regression and hold-out validation. We have employed a threshold mechanism to manipulate with the attendance matrix, and an AUC statistical indicator. Self-adaptive differential evolution was used as optimization algorithm. Here, the general goal is to find out whether FS can contribute to reduce the number of marketing phone calls and thus decrease marketing costs. We employ a custom data reduction algorithm for feature selection, and compare its performance to performance of: (1) complete dataset and (2) recursive feature elimination (RFE) method. We combine the favourable modeling characteristics of an artificial neural network and valuable optimization characteristics of a self-adaptive differential evolution. As a benchmark, we adopt a known UCI Machine Learning dataset named "Bank Marketing Data Set" [26], composed by Portuguese banking institutions during a campaign of phone calls. The dataset consists of many features (which we even enlarge) that are divided into three major groups: personal, social and financial. The purpose of the dataset is to predict whether a client is willing to make (subscript) a bank deposit or not. The problem we are facing is building a model from the dataset in a way that would maximize prediction performance. An example of bank marketing classification is shown in [32]. Here, we propose an automatic FS optimization process to obtain: (1) simpler modeling problem, (2) higher prediction performance, and (3) lower time complexity. Unlikely to [14], here we evaluate trial solutions using the ten-fold cross-validation procedure. By cross-validating, we avoid any random (bias) effects, which are present if the single hold-out validation is used, and thus expect that the proposed solution will actually increase prediction performance and lower the marketing costs simultaneously.

The outline of the paper is as follows. Section 2 deals with the basic information needed for understanding subjects that follows. Section 3 outlines the proposed method, while Sect. 4 presents the obtained results. Section 5 concludes the paper and outlines directions for the future work.

2   Related Information

This section focuses on the background information that is needed for understanding the subjects that follow. In line with this, two prerequisites are necessary, i.e. an artificial neural networks and a differential evolution. Additionally to the original differential evolution, its self-adaptive variant is also illustrated. In the remainder of the paper, the aforementioned topics are presented in detail.

2.1   Artificial Neural Network

An artificial neural network (ANN) is a common ML modelling tool, known for its universal versatility for arbitrary approximations tasks [12]. Typically, ANN consists of multiple layers of perceptrons, i.e. building blocks, which simulate the behaviour of human. A perceptron consists of bias (sum), weight (scaling) and transfer function. Many perceptrons connected into the ANN can address diverse types of problems, such as regression, classification, clustering, dimensionality reduction and others [20].

A two-phase process is typically employed in order to evaluate the ANN: ANN-training (learning) and ANN-validation (evaluation). Those are run sequentially on two non-overlapping samples of the original dataset, i.e. training and validation samples. At first, the training is performed and then validation is executed to derive the predictive ability of the ANN by directly comparing the known and predicted results. However, such inference may be biased, due to the lack of generalization or representativeness of the validation dataset. In order to avoid this problem, a special type of validation is used, i.e. k-fold cross validation. The latter splits the dataset into k equal folds, and uses each of them exactly once for the validation, and the rest of the $k - 1$ folds for training [12]. In this way, multiple (k) training and validation processes are run to obtain k results, which can then be averaged to form a consistent and unbiased measure of predictive ability.

2.2   Differential Evolution

Differential Evolution is an evolutionary, population-based, nature-inspired algorithm for global optimization, proposed by Storn and Price [36]. It belongs to a family of Evolutionary Algorithms (EA). Like other stochastic population-based nature-inspired algorithms, DE represents its candidate solutions as D-dimensional real-valued vectors $\mathbf{x}_i^{(t)}$ with elements $x \in [0, 1]$, in other words [34, 37]:

$$\mathbf{x}_i^{(t)} = \{x_{i,1}^{(t)}, \ldots, x_{i,D}^{(t)}\}, \quad \text{for } i = 1, \ldots, NP, \tag{1}$$

where NP denotes the population size, and D is the dimension of the problem.

The principle of the algorithm bases on applying the three genetic operators: mutation, crossover and selection, launched sequentially. The first among them, i.e. mutation, is used to encourage genetic diversity of the solutions, and prevent convergence to the local optimum. It is performed by taking the difference vector between two individuals and scaling its difference to form the mutant vector. For instance, mutation strategy, called 'DE/rand/1/bin', is formalized in Eq. (2):

$$\mathbf{u}_i^{(t)} = \mathbf{x}_{r0}^{(t)} + F \cdot (\mathbf{x}_{r1}^{(t)} - \mathbf{x}_{r2}^{(t)}), \quad \text{for } i = 1, \ldots, NP, \tag{2}$$

where $F \in (0.0, 1.0]$ represents the (stepsize) scaling factor and NP the population size. Thus, indices $r0$, $r1$, and $r2$, are the randomly selected numbers, drawn from uniform distribution in the interval $1, \ldots, NP$. Those denote the corresponding solution that must be different from the target. Crossover is then employed to combine the mutant vector with individuals $\mathbf{x}_{ri}$ to form the trial vector w_i, represented by Eq. (3):

$$w_{i,j}^{(t)} = \begin{cases} u_{i,j}^{(t)}, & \text{if } \text{rand}_j(0,1) \leq CR \vee j = j_{rand}, \\ x_{i,j}^{(t)}, & \text{otherwise}, \end{cases} \tag{3}$$

where $CR \in [0.0, 1.0]$ means the crossover rate and $j = 1, \ldots, D$. The third genetic operator, i.e., selection, is used to compare the two vectors, i.e., trial $\mathbf{w}_i$ and target $\mathbf{x}_i$, and select the better between them. The selection procedure is outlined in Eq. (4):

$$\mathbf{x}_i^{(t+1)} = \begin{cases} \mathbf{w}_i^{(t)}, & \text{if } f(\mathbf{w}_i^{(t)}) \leq f(\mathbf{x}_i^{(t)}), \\ \mathbf{x}_i^{(t)}, & \text{otherwise}. \end{cases} \tag{4}$$

The better among trial and target vector is preserved into the candidate solution vector $\mathbf{x}_i$ which proceeds into the next generation.

Self-adaptive Differential Evolution. DE is a simple and very useful algorithm, which can be hybridized to improve its search performance. For example, Brest et al. propose the self-adapted version of DE (jDE) [6] that self-adapts control parameters F and CR during the search process in order to simulate the "evolution of the evolution". In case of jDE, representation of individuals changes according to Eq. (5):

$$\mathbf{x}_i^{(t)} = (x_{i,1}^{(t)}, x_{i,2}^{(t)}, \ldots, x_{i,M}^{(t)}, F_i^{(t)}, CR_i^{(t)}), \tag{5}$$

where control parameters $F_i^{(t)}$ and $CR_i^{(t)}$ undergo specific variation operator, formalized in Eqs. (6) and (7):

$$F_i^{(t+1)} = \begin{cases} F_l + \text{rand}_1 * (F_u - F_l) & \text{if } \text{rand}_2 < \tau_1, \\ F_i^{(t)} & \text{otherwise}, \end{cases} \tag{6}$$

and

$$CR_i^{(t+1)} = \begin{cases} \text{rand}_3 & \text{if rand}_4 < \tau_2, \\ CR_i^{(t)} & \text{otherwise}. \end{cases} \tag{7}$$

Here, $\text{rand}_{i=1\ldots4} \in [0,1]$ represent the random numbers drawn from uniform distribution in interval $[0,1]$, τ_1, τ_2 learning rates and F_l, F_u the lower and upper bounds for scale factor F.

3 Proposed Method

This section presents the practical implementation of the KDD procedure and the synthesis of the proposed FS. As mentioned in the Introduction, this procedure consists of six phases. In this sense, the aims of the phases are discussed in a nutshell.

The first among them, i.e., problem specification, deals with the purpose of the study. Here, the provided database is examined, and the final objectives of the study are considered. The hypotheses to be checked are considered, and the expected results discussed with the help of expert knowledge. Also, relevant information is explored about the problem and related literature.

By completing the problem specification phase, the problem understanding phase follows. Here, the data and any interconnections among the data are checked visually to obtain basic data comprehension. After that, features and response variable(s) are selected to form a reduced representation of the database, i.e. dataset [13]. For continuing the study, only the dataset is relevant.

The third phase of KDD is the data preprocessing. Two sub-phases are contained here, i.e. data preparation and data reduction. During the former sub-phase, tasks like data cleaning, integration, normalization and transformation are applied. During the latter, the original database is reduced, to eliminate the redundancy or irrelevancy of some of the elements [15]. Here, four methods exist: (1) feature selection (FS), (2) instance selection (IS), (3) discretization and (4) feature/instance extraction. FS deals with eliminating the explanatory variables from the dataset in order to eliminate their redundancy, while IS deals with eliminating the instances from the dataset. The discretization is a data reduction principle, where the domain is simplified into discrete regions. Feature/Instance extraction is a method where new variables are generated.

The fourth KDD phase is the data mining. It is the central and critical phase of the KDD, where actual hidden patterns, relations and interconnections are searched for. First, the appropriate DM task and the DM method are selected. Typically, DM tasks are: regression, classification, clustering and others. The more commonly DM methods are: ANNs, decision trees, support vector machines (SVM), rule-based algorithms and others. It is known that each DM method does not suit each problem well, and that some experimenting might sometimes be necessary.

The fifth KDD phase is the evaluation. Here, the researcher interprets the performance of the KDD. Typically, performance is evaluated by the model's performance using a single criterion or multiple selection criteria. Many selection criteria exist, such as information, distance, dependence, consistency and accuracy measures. Each of them defines a critical evaluation criterion - performance (ability) of the DM.

The last step of KDD is the result exploitation, where the researcher uses the knowledge discovered practically. Visualization, documentation and reporting come into play here, together with comparison to the expectations set in the first phase. Also, built knowledge discovered can be used pro-actively in the daily routine. In the subsections that follow, the aforementioned KDD phases are illustrated from the proposed method point of view in detail.

3.1   Problem Specification

The purpose of our study is to identify and select the best subset, which would assure to find the maximal number of bank clients, willing to subscript the bank deposit. We employ the dataset accessible at UCI Machine Learning, i.e. the Bank Marketing Dataset that was collected by Moro, Cortez and Rita in 2014 [26]. The dataset comes from Portuguese banking institutions which have been advertising their products. Using campaign phone calls, they have been contacting clients and promoting bank deposit subscriptions. The institutions have also, simultaneously, been recognizing the client's personal, social and financial habits, and client's decision to subscript the deposit or not [33].

We are sure that the KDD from this dataset could help us predict interested clients. Banking institutions can nevertheless save enormous efforts and costs of bank marketing if they avoid contacting each client individually again and again. In order to decrease the pool of potential bank depositors, we propose to build a model with which clients with past information could be evaluated, and only most potential among them would be contacted in future.

3.2   Problem Understanding

In the second KDD phase, we discuss briefly the features held in the dataset to obtain basic comprehension of data. Actually, collected data from an original dataset have been accumulated into a table to form an original dataset $\mathbf{X}$, which are depicted in Table 1. Besides features, a response variable *deposit_subscription* is added to the table also. The *deposit_subscription* is a binary value with the following meaning: 0 means reject, and 1 means the subscription of the bank deposit.

Let us mention that the original dataset $\mathbf{X}$ consists of 20 features and 41,188 instances describing clients. Although this seems reasonable, much less than 41,188 clients are examined practically, since the majority of them are contacted more than once during the campaign period.

As can be seen from the Table 1, the features can be either numerical or categorical. Each numerical variable is defined with a corresponding range of feasible values. On the other hand, the categorical variables are specified with a set of discrete values that are also presented in the table.

Table 1. List of explanatory and dependent variables in the original dataset **X**.

No	Feature	Type of feature	Range of feature
1	age	Numerical	17 – 98 years
2	job	Categorical	Administrator, blue-collar, entrepreneur, housemaid, management, retired, self-employed, services, student, technician, unemployed, unknown
3	marital	Categorical	Divorced, married, single, unknown
4	education	Categorical	basic.4y, basic.6y, basic.9y, high.school, illiterate, professional.course, university.degree, unknown
5	default	Categorical	No, yes, unknown
6	housing	Categorical	No, yes, unknown
7	loan	Categorical	No, yes, unknown
8	contact	Categorical	Cellular, telephone
9	month	Categorical	March, April, May, June, July, August, September, October, November, December
10	day_of_week	Categorical	Monday, Tuesday, Wednesday, Thursday, Friday
11	duration	Numerical	0 – 4918
12	campaign	Numerical	1 – 56
13	pdays	Numerical	0 – 999
14	previous	Numerical	0 – 27
15	poutcome	Categorical	Failure, success, nonexistent
16	emp.var.rate	Numerical	−3.4 – 1.4
17	cons.price.idx	Numerical	92.201 – 94.767
18	cons.conf.idx	Numerical	−50.8 –26.9
19	euribor3m	Numerical	0.634 – 5.045
20	nr.employed	Numerical	4963.6 – 5228.1
21	deposit_subscription	Binary	0 – 1

3.3   Data Preprocessing

In our study, data preparation consists of two data transformations: dummification, and normalization. Dummification is a transformation, where categorical or numerical variables are transformed into binary features using one-hot encoding. One-hot encoding is a transformation that encodes categorical variables with multiple classes into a binary vector. Exactly one of the binary vector values is 1 and the rest of them are 0. On the other hand, normalization stands for transformation, which modifies the values to the interval [0,1] proportionally.

Data Preparation. The explanatory variables are dummified and normalized as presented in Table 2, where *pdays* although it is numerical, is modified using the dummification. Value 0 presents the situation that the client has never been contacted before, and vice-versa. In this way, the original dataset is widened to 70 variables that form the adjusted dataset **X′** [14]. Actually, 62 binary variables and 8 numerical normalized variables are included here. All of them are presented in the correlation analysis plot (Fig. 1). High correlation and, thus, redundancy is reported for the *age*, *months*, *day* and *marital* binary variables.

Table 2. Transformation of categorical and numerical variables.

Categorical transformations		Numerical transformations	
age	8 dummies	*Duration*	[0, 1] normalization
job	12 dummies	*Campaign*	[0, 1] normalization
marital	4 dummies	*Previous*	[0, 1] normalization
education	8 dummies	*emp.var.rate*	[0, 1] normalization
default	3 dummies	*cons.price.idx*	[0, 1] normalization
housing	3 dummies	*cons.conf.idx*	[0, 1] normalization
loan	3 dummies	*euribor3m*	[0, 1] normalization
contact	2 dummies	*nr.employed*	[0, 1] normalization
month	10 dummies		
day_of_week	5 dummies		
pdays	1 dummy		
poutcome	3 dummies		

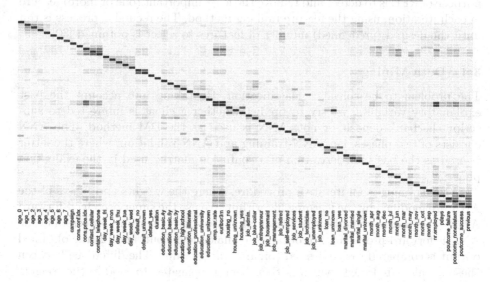

Fig. 1. Correlation analysis plot of adjusted dataset $\mathbf{X}'$.

In addition, high correlation is examined among the majority of numerical variables and for *months* binary variables. Almost perfect correlation is observed between variables *emp.var.rate* and *euribor3m*.

Data Reduction. The proposed wrapper-based FS is implemented using the modified WFS-jDE that is obtained from the original one by modifying the following jDE elements: (1) representation of individuals, and (2) fitness function evaluation. While the representation of individuals is performed according to the

Eq. (1), the fitness function evaluation is calculated as follows: Let us assume that an attendance vector $\mathbf{a}_i^{(t)} = \{a_{i,j}\}$, corresponding to each candidate solution $\mathbf{x}_i^{(t)}$, is given. The elements of the attendance vector are calculated according to Eq. (8), as follows:

$$\mathbf{a}_i^{(t)} = \begin{cases} 0, \text{ if } \mathbf{x}_i \leq 0.5 \\ 1, \text{ otherwise,} \end{cases} \tag{8}$$

Then, the attendance vector $\mathbf{a}_i^{(t)}$ specifies the presence or absence of specific features in the reduced subset $\mathbf{X}^* \subset \mathbf{X}'$.

$$g : \mathbf{X}' \xrightarrow{\mathbf{a}_i} \mathbf{X}^* \tag{9}$$

where the subset $\mathbf{X}^*$ contains features with an attendance vector value of 1, and omits the features with an attendance vector value of 0. Each subset $\mathbf{X}^*$ is evaluated using the CV, which will be described in Subsect. 3.5.

Alternative to the described FS method is the recursive feature elimination algorithm, or shortly RFE. Here, logistic regression is used as an estimator. The purpose of RFE is to detect and remove the least important (one or more) feature in each iteration, using the feature ranking method. The recursive process is run until the desired (predefined) number of features to select is obtained [30].

3.4 Data Mining

The problem to be solved is classification. By taking into account the past explanatory variables, we try to predict whether a client is interested to subscript the bank deposit or not. ANN is used as the DM method. The ANN consists of two phases, i.e., ANN-training and ANN-validation, where the latter is used as the basis for derivation of a confusion matrix, used in the evaluation phase.

ANN-training is an iterative procedure, where the weights and biases of the ANN are modified by the learning algorithm. We use the supervised feed-forward ANN. This means that the ANN-training is two-phase learning process: (1) the ANN is forward-propagated to obtain the analogue output, and (2) the obtained output is compared to the desired (binary) target value. The difference between these is calculated next, which is then back-propagated to modify the weights and biases of the ANN.

ANN-validation consists of the forward-propagation only. Here, the validation data not seen during the training are used. Each forward-propagation gives a classification output that is compared with the actual class (client subscripts/rejects the deposit). This forms four different scenarios, that are summarized in the confusion matrix. ANN-validation is due to the 10-fold cross-validation (CV) mechanism performed 10 times.

3.5 Evaluation

Performance of a classifier is evaluated in the evaluation phase of KDD. Evaluation of the classification results is performed on the basis calculated by the

Table 3. An example of a confusion matrix.

	True YES	True NO
Predicted YES	TP	FP (type I error)
Predicted NO	FN (type II error)	TN

confusion matrix. Selected subset $\mathbf{X}^*$ from the data reduction sub-phase is next split into fixed and non-overlapping $k = 10$ folds for the CV calculations. Nine of them are used for ANN-training, while the tenth for the ANN-validation. This process is repeated until all of the folds are used exactly once for ANN-validation. Prediction results are recorded afterwards, and saved in the form of confusion matrix (Table 3), i.e. a tabular indicator of a goodness-of-classification, consisting of four quadrants, and differentiating between true and false predictions. In a confusion matrix, proper prediction of a deposit subscription means a true positive prediction, while proper prediction of rejection a true negative. Two other cases exist: A false positive for an improper prediction of rejection (type I error), and a false negative for the improper prediction of subscription (type II error). Since the confusion matrix is a universal statistical indicator of prediction (classification) performance, many derivative measures of accuracy can be calculated from it. Specifically, we use the confusion matrix for the calculation of overall accuracy and sensitivity, which can be outlined by Eqs. (10) and (11).

$$accuracy = \frac{TP+TN}{TP+FP+FN+TN},\tag{10}$$

$$sensitivity = \frac{TP}{TP+FN},\tag{11}$$

where both the *accuracy* and *sensitivity* are used for a fitness function computation, as:

$$ff(trial\_solution) = 0.5 \cdot \overline{accuracy} + 0.5 \cdot \overline{sensitivity}.\tag{12}$$

Variables $\overline{accuracy}$ and $\overline{sensitivity}$ present the mean (average) values of accuracy and sensitivity over ten folds, used in CV. We are dealing with an imbalanced dataset, and the accuracy may solely be a biased measure. In order to prevent it, a combination is used with sensitivity. By taking the coefficients of each measure 0.5 we treat both of them equally. The obtained fitness function is next fed as a feedback loop into the optimization algorithm, and the best candidate solution with the best subset $\mathbf{X}^{(best)}$ can be extracted, as follows:

$$\mathbf{X}^{(best)} = max\left(f(\mathbf{X}^*)\right)\tag{13}$$

Results are exploited on the basis of the obtained $\mathbf{X}^{(best)}$.

3.6   Result Exploitation

The obtained result (built model) can be used on a regular basis to classify interested banking clients. The complete WFS-jDE procedure is shown in the use-case diagram in Fig. 2.

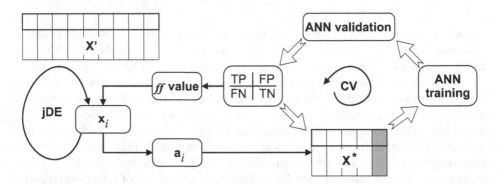

Fig. 2. Use-case diagram of WFS-jDE. Two loops are shown, i.e. a jDE loop, which is controlled by a maximum number of evaluations ($nFES$), and a CV loop, that presents the $k = 10$-fold cross-validation. First, candidate solution vector $\mathbf{x}_i$ is generated, and transformed according to attendance vector $\mathbf{a}_i$ to form a trial subset $\mathbf{X}^*$. This is then trained sequentially, and validated using the CV mechanism, and the results are visualized in a confusion matrix individually for each fold. Fitness function value is calculated once all the folds are evaluated.

4   Experiments and Results

The purpose of the following section is twofold. First, we investigate if we can predict bank clients interested in making deposits. Second, we examine if the proposed WFS-jDE can be used for improving prediction performance.

For the experiments, the widened Bank Marketing Data Set (adjusted dataset $\mathbf{X}'$) introduced in subsection 3.3 was used. ANN was used as a modeling tool to ensure the confusion matrix calculation and jDE as an optimization algorithm to optimize the attendance vector.

ANN was implemented in the Python programming language using, the Keras deep learning library [9]. The following configuration of ANN was used:

- input layer with number of input neurons N_i varying by the number of features,
- first hidden layer with a number of hidden neurons $N_{h1} = 20$,
- second hidden layer with a number of hidden neurons $N_{h2} = 20$,
- output layer with a single output neuron $N_o = 1$.

Table 4. Configuration of ANN.

Parameter	Value
Number of epochs N_{epochs}	10
Batch size $batch\_size$	32
Learning constant lr	0.001
Loss function	$binary\_crossentropy$

A sigmoid function is used in each layer. The architecture of ANN was taken from the paper [11], which proposes 48 input neurons, 20 neurons in the first hidden layer, 15 neurons in the second hidden layer, and a single output neuron. An advanced ANN training algorithm Adam [16] was used with fixed size of epochs. No early-stopping criterion was used. Table 4 presents ANN parameter configuration. The initial configuration of the jDE optimization algorithm is referenced in the Table 5. Scikit-learn library was taken to provide RFE procedure [29]. RFE is the alternative to the proposed solution, which requires the setting of the final number of features. An adjusted dataset $\mathbf{X}'$ is used as a basis. In what follows, two experimental tests are conducted for the adjusted dataset $\mathbf{X}'$, best subset $\mathbf{X}^{(\mathbf{best})}$ and RFE subset $\mathbf{X}^{(\mathbf{RFE})}$:

- basic evaluation of prediction performance,
- detailed comparison among statistical indicators.

Table 5. Initial configuration of jDE optimization algorithm.

Parameter	Value
Initial scaling factor F	0.5
Initial crossover ratio CR	0.9
Self-adaptive learning rate τ_1	0.1
Self-adaptive learning rate τ_2	0.1
Population size NP	100
Number of evaluations $nFES$	10,000

First, basic evaluation of prediction performance is presented. Second, detailed comparison among statistical indicators is displayed. Confusion matrix is used for basic evaluation, while, for detailed evaluation, statistical indicators (measures of accuracy) are derived from the confusion matrix. Following eight statistical indicators are used: accuracy, sensitivity, specificity, precision, negative predictive value (NPV), type I and type II errors, and F1 score. Additionally, fitness function value (ff) is displayed for comparison. Since the CV mechanism is employed, average values are displayed for all of the obtained results. In the remainder of the paper, the aforementioned experiments are explained in detail.

Table 6. Average confusion matrix for adjusted dataset $\mathbf{X}'$.

	True YES	True NO
Predicted YES	216.60	122.00
Predicted NO	247.40	3532.80

Table 7. Average confusion matrix for best subset $\mathbf{X}^{(\text{best})}$.

	True YES	True NO
Predicted YES	234.60	141.00
Predicted NO	229.40	3513.80

4.1 Basic Evaluation

Basic evaluation is performed for the adjusted dataset $\mathbf{X}'$, best subset $\mathbf{X}^{(\text{best})}$ and RFE subset $\mathbf{X}^{(\text{RFE})}$. The results of the adjusted dataset are obtained by running the classification on the complete adjusted dataset $\mathbf{X}'$. The average confusion matrix is recorded in Table 6, which exhibits basic prediction performance. In average, 216.6 clients are predicted properly for bank deposit subscription and 3532.8 clients on average for bank deposit rejection. 247.4 clients are predicted incorrectly, since they actually wish to subscript a deposit, but the model predicts inversely. 122 clients are predicted incorrectly as well, since the model predicts them to subscribe, but they actually reject the bank deposit. Table 6 is thus a benchmark predictive ability, which we would try to improve using the WFS-jDE procedure. Table 7 exhibits results on the best subset $\mathbf{X}^{(\text{best})}$, which are obtained by applying the iterative WFS-jDE procedure on the complete dataset $\mathbf{X}'$. The dimension is reduced from 70 features to 37, meaning that almost half of the features are omitted. By applying the WFS-jDE, the number of properly predicted subscriptions rises to 234.6, compared to 216.6 in Table 6, and the number of properly predicted rejections diminishes from 3532.8 to 3513.8. Although the first characteristic is very welcome - the rise of proper predictions provides more bank deposit subscriptions, this rise is conditioned by a fall of proper rejection predictions. The type I error increases slightly due to that reason, and the type II error decreases. The next subsection quantifies these effects. Categorical variables that are applicable in the best subset $\mathbf{X}^{(\text{best})}$ (those that are not omitted), are outlined in the Table 8. Alongside, the following binary and numerical variables are present in the best subset $\mathbf{X}^{(\text{best})}$: *pdays, duration, emp.var.rate, cons.price.idx, cons.conf.idx* and *nr.employed.*

Table 8. List of categorical variables in the best subset $\mathbf{X}^{(best)}$.

Features	Feature values
age	17–25, 26–34, 44–52, 53–61, 62–70
job	Management, services, technician
marital	Unknown
education	basic.4y, basic.9y, professional.course, university.degree, unknown
month	March, May, June, September, October, December
day_of_week	Wednesday, Thursday
default	Yes
housing	No, yes, unknown
loan	Yes
poutcome	Failure, success, nonexistent
contact	Telephone

Listed variables in the best subset $\mathbf{X}^{(best)}$ are similar to two FS studies. Both of the studies [11,28] rank the most important feature to be *duration*. [28], who use the information gain and chi-square characteristic, then rank *poutcome*, *month*, *pdays* and *contact*. On the other hand, according to the sensitivity analysis, ranks *month*, *poutcome*, *contact* and *job* to be the other most important features [11]. Third evaluation is the evaluation of the feature selection alternative - RFE method with the number of variables to select set to 35. According to the confusion matrix Table 9, this alternative is the worst among all. True positive instance is decreased drastically, scoring barely 85 clients. Number of false positives decreases as well, which is desired, but the false negatives increase. True negative instance scores the highest value among all. Table 10 lists the categorical variables, used in the RFE subset $\mathbf{X}^{(RFE)}$. It is noticeable that each feature that is listed in the table, incorporates all of its feature values (except the *day_of_week* feature, where the Tuesday and Wednesday feature values are missing).

Table 9. Average confusion matrix for RFE subset $\mathbf{X}^{(RFE)}$.

	True YES	True NO
Predicted YES	85.30	44.40
Predicted NO	378.70	3610.40

Table 10. List of categorical variables in the RFE subset $\mathbf{X}^{(\mathbf{RFE})}$.

Features	Feature values
age	17–25, 26–34, 35–43, 44–52, 53–61, 62–70, 71–79, 80+
job	Administrator, blue-collar, entrepreneur, housemaid, Management, retired, self-employed, services, Student, technician, unemployed, unknown
marital	Single, married, divorced, unknown
day_of_week	Monday, Thursday, Friday
default	No, yes, unknown
poutcome	Failure, success, nonexistent
contact	Telephone, cellular

4.2   Detailed Evaluation

The aim of this section is to provide the detailed evaluation between the adjusted dataset $\mathbf{X}'$, best subset $\mathbf{X}^{(\mathbf{best})}$ and RFE subset $\mathbf{X}^{(\mathbf{RFE})}$. The results displayed in Table 11 present the mentioned statistical indicators, where NPV means negative predictive value and *ff* means fitness function value. The displayed statistical indicators are averaged for the ten folds used in the CV mechanism. Examination of the detailed results identifies a slight drop in overall accuracy by applying the proposed WFS-jDE solution. Substantial increase in sensitivity is indicated, which causes fitness function value *ff* to increase as well. An increase of fitness function value indicate the successfulness of the WFS-jDE. Specificity is due to the drop of proper rejection predictions lowered marginally compared to adjusted dataset, and thus increases type I errors proportionately. Specificity is the highest for the RFE subset. On the other hand, type II error lowers drastically which

Table 11. Detailed evaluation of prediction results.

Avg. indicator	Adjusted dataset $\mathbf{X}'$	Best subset $\mathbf{X}^{(\mathbf{best})}$	RFE subset $\mathbf{X}^{(\mathbf{RFE})}$
No. of features	70	37	35
Accuracy	0.9103	0.9101	0.8973
Sensitivity	0.4668	0.5056	0.1838
Specificity	0.9666	0.9614	0.9879
Precision	0.6407	0.6272	0.6572
NPV	0.9346	0.9388	0.9051
Type I error	0.0334	0.0386	0.0121
Type II error	0.5332	0.4944	0.8162
F1 score	0.5381	0.5584	0.2869
ff value	0.6886	0.7078	0.5406

causes that far less clients are "ignored". After all, regardless of any statistical indicator, the number of bank deposit subscriptions increases strongly. The RFE procedure decreases the overall accuracy and drastically decreases the sensitivity. Due to such a low value of sensitivity and consequently fitness function value, this FS procedure is not appropriate for our problem. A good quality of the WFS-jDE comes in the slight increase of type I errors but significant drop of type II errors. The dataset is imbalanced and thus biased in some way. A special treatment is necessary, which comes in the form of fitness function. The latter tries to take the imbalance into account at least a bit. Different fitness functions would establish different results.

The results can be compared to those obtained in [14]. Type II error there is remarkably lower, i.e. 27 for the "original database" and 28 for the "reduced database". The authors report that their accuracy increases by almost 2% by employing the FS, while we experienced a drop in accuracy. However, the authors there employed a simple hold-out validation and a logistic regression. Additionally, they implemented a special procedure for threshold (TH), which was used for mapping candidate solutions. They used simultaneous optimization of threshold TH during the optimization process and thus improved the reaction of FS significantly on the imbalanced dataset. Instead of custom fitness function from accuracy and sensitivity, they optimized the AUC score. All those changes call for a difficult comparison between the [14] and our study. Another study that deals with the bank marketing is [28]. The authors here compared the F1 score. By employing several traditional FS applications, the authors show that maximal increase of F1 score is 0.01. In our case, an increase of more than 0.02 is applicable.

5 Conclusion

In this paper, we have proposed a wrapper-based feature selection using the ANN and jDE. We have tested its efficiency on a Bank Marketing Data Set. We figured out that modeling the past information about bank clients is a suitable task that might come handy on an everyday basis.

We have shown that the proposed FS procedure reduces the size of the dataset successfully. From an initial 70 variables, we formed the best subset of 37 variables, i.e. almost half less variables. Such a decrease of the number of variables not only reduces the complexity of the dataset, but also improves the predictive performance of a classifier. The RFE alternative, which is more common in practice, caused the predictive ability to suffer. Although, we can say that on this case study, FS procedure can be used beneficially to improve the predictive ability by eliminating some redundant or irrelevant features.

The proposed WFS-jDE is especially suitable for imbalanced datasets, due to the arbitrary selection of fitness function. However, self-adaptation of the threshold TH is highly desired in such cases.

In future, we would like to test WFS-jDE with the self-adaptation of TH on several diverse datasets. We would like to employ universal fitness function, and run the jDE with a higher number of evaluations, to ensure the convergence. Constrained optimization should be utilized as well, to control the number of features better in the reduced dataset.

References

1. Aggarwal, C.C.: Data Classification: Algorithms and Applications. CRC Press, Boca Raton (2014)
2. Al-Tashi, Q., Kadir, S.J.A., Rais, H.M., Mirjalili, S., Alhussian, H.: Binary optimization using hybrid grey wolf optimization for feature selection. IEEE Access **7**, 39496–39508 (2019)
3. Apolloni, J., Leguizamón, G., Alba, E.: Two hybrid wrapper-filter feature selection algorithms applied to high-dimensional microarray experiments. Appl. Soft Comput. **38**, 922–932 (2016)
4. Bates, D.W., Saria, S., Ohno-Machado, L., Shah, A., Escobar, G.: Big data in health care: using analytics to identify and manage high-risk and high-cost patients. Health Aff. **33**(7), 1123–1131 (2014)
5. Bellman, R.: Adaptive Control Processes: A Guided Tour Princeton University Press. Princeton, New Jersey (1961)
6. Brest, J., Greiner, S., Boškovič, B., Mernik, M., Žumer, V.: Self-adapting control parameters in differential evolution: a comparative study on numerical benchmark problems. IEEE Trans. Evol. Comput. **10**(6), 646–657 (2006)
7. Cardona, L., Moreno, L.A.: Cash management cost reduction using data mining to forecast cash demand and LP to optimize resources. Memetic Comput. **4**(2), 127–134 (2012)
8. Chandrashekar, G., Sahin, F.: A survey on feature selection methods. Comput. Electr. Eng. **40**(1), 16–28 (2014)
9. Chollet, F., et al.: Keras (2015). https://keras.io
10. Da Cunha, C., Agard, B., Kusiak, A.: Selection of modules for mass customisation. Int. J. Prod. Res. **48**(5), 1439–1454 (2010)
11. Elsalamony, H.A., Elsayad, A.M.: Bank direct marketing based on neural network. Int. J. Eng. Adv. Technol. (IJEAT) **2**(6) (2013)
12. Ertel, W.: Introduction to Artificial Intelligence. Springer, Heidelberg (2018)
13. Fayyad, U., Piatetsky-Shapiro, G., Smyth, P.: From data mining to knowledge discovery in databases. AI Mag. **17**(3), 37–37 (1996)
14. Fister, D., Fister, I., Jagrič, T., Fister Jr, I., Brest, J.: A novel self-adaptive differential evolution for feature selection using threshold mechanism. In: 2018 IEEE Symposium Series on Computational Intelligence (SSCI), pp. 17–24. IEEE (2018)
15. García, S., Luengo, J., Herrera, F.: Data Preprocessing in Data Mining. Springer, New York (2015). https://doi.org/10.1007/978-3-319-10247-4
16. Kingma, D.P., Ba, J.: Adam: a method for stochastic optimization. arXiv preprint arXiv:1412.6980 (2014)
17. Kohavi, R., John, G.H.: Wrappers for feature subset selection. Artif. Intell. **97**(1–2), 273–324 (1997)
18. Labani, M., Moradi, P., Ahmadizar, F., Jalili, M.: A novel multivariate filter method for feature selection in text classification problems. Eng. Appl. Artif. Intell. **70**, 25–37 (2018)
19. Lasi, H., Fettke, P., Kemper, H.-G., Feld, T., Hoffmann, M.: Industry 4.0. Bus. Inf. Syst. Eng. **6**(4), 239–242 (2014)

20. LeCun, Y., Bengio, Y., Hinton, G.: Deep learning. Nature **521**(7553), 436 (2015)
21. Liu, H., Zhou, M.C., Liu, Q.: An embedded feature selection method for imbalanced data classification. IEEE/CAA J. Automatica Sinica **6**(3), 703–715 (2019)
22. Liu, Z.-Z., Huang, J.-W., Wang, Y., Cao, D.-S.: ECoFFeS: a software using evolutionary computation for feature selection in drug discovery. IEEE Access **6**, 20950–20963 (2018)
23. Meng, L.: Embedded feature selection accounting for unknown data heterogeneity. Expert Syst. Appl. **119**, 350–361 (2019)
24. Mafarja, M., Mirjalili, S.: Whale optimization approaches for wrapper feature selection. Appl. Soft Comput. **62**, 441–453 (2018)
25. Mallik, P., Roy, C., Maheshwari, E., Pandey, M., Rautray, S.: Analyzing student performance using data mining. In: Hu, Y.-C., Tiwari, S., Mishra, K.K., Trivedi, M.C. (eds.) Ambient Communications and Computer Systems. AISC, vol. 904, pp. 307–318. Springer, Singapore (2019). https://doi.org/10.1007/978-981-13-5934-7_28
26. Moro, S., Cortez, P., Rita, P.: A data-driven approach to predict the success of bank telemarketing. Decis. Support Syst. **62**, 22–31 (2014)
27. Osanaiye, O., Cai, H., Choo, K.-K.R., Dehghantanha, A., Xu, Z., Dlodlo, M.: Ensemble-based multi-filter feature selection method for DDOS detection in cloud computing. EURASIP J. Wirel. Commun. Netw. **1**, 130 (2016)
28. Parlar, T., Acaravci, S.K.: Using data mining techniques for detecting the important features of the bank direct marketing data. Int. J. Econ. Fin. Issues **7**(2), 692–696 (2017)
29. Pedregosa, F., et al.: Scikit-learn: machine learning in Python. J. Mach. Learn. Res. **12**, 2825–2830 (2011)
30. Pullanagari, R., Kereszturi, G., Yule, I.: Integrating airborne hyperspectral, topographic, and soil data for estimating pasture quality using recursive feature elimination with random forest regression. Rem. Sens. **10**(7), 1117 (2018)
31. Ramjee, S., Gamal, A.E.: Efficient wrapper feature selection using autoencoder and model based elimination. arXiv preprint arXiv:1905.11592 (2019)
32. Scherer, M., Smolag, J., Gaweda, A.: Predicting success of bank direct marketing by neuro-fuzzy systems. In: Rutkowski, L., Korytkowski, M., Scherer, R., Tadeusiewicz, R., Zadeh, L.A., Zurada, J.M. (eds.) ICAISC 2016. LNCS (LNAI), vol. 9693, pp. 570–576. Springer, Cham (2016). https://doi.org/10.1007/978-3-319-39384-1_50
33. Serrano-Silva, Y.O., Villuendas-Rey, Y., Yáñez-Márquez, C.: Automatic feature weighting for improving financial decision support systems. Decis. Support Syst. **107**, 78–87 (2018)
34. Simon, D.: Evolutionary Optimization Algorithms. Wiley, Hoboken (2013)
35. Srinivasan, U., Arunasalam, B.: Leveraging big data analytics to reduce healthcare costs. IT Prof. **15**(6), 21–28 (2013)
36. Storn, R., Price, K.: Differential evolution-a simple and efficient heuristic for global optimization over continuous spaces. J. Glob. Optim. **11**(4), 341–359 (1997)
37. Viktorin, A., Senkerik, R., Pluhacek, M., Kadavy, T., Zamuda, A.: Distance based parameter adaptation for success-history based differential evolution. Swarm Evol. Comput. **50**, 100462 (2018)
38. Wang, S., Tang, J., Liu, H.: Embedded unsupervised feature selection. In: Twenty-Ninth AAAI Conference on Artificial Intelligence (2015)
39. Wang, Y., Huang, J.-J., Zhou, N., Cao, D.-S., Dong, J., Li, H.-X.: Incorporating PLS model information into particle swarm optimization for descriptor selection in QSAR/QSPR. J. Chemom. **29**(12), 627–636 (2015)

40. Xindong, W., Zhu, X., Gong-Qing, W., Ding, W.: Data mining with big data. IEEE Trans. Knowl. Data Eng. **26**(1), 97–107 (2014)
41. Xue, B., Zhang, M., Browne, W.N., Yao, X.: A survey on evolutionary computation approaches to feature selection. IEEE Trans. Evol. Comput. **20**(4), 606–626 (2016)
42. Yu, L., Liu, H.: Feature selection for high-dimensional data: a fast correlation-based filter solution. In: Proceedings of the 20th International Conference on Machine Learning (ICML-03), pp. 856–863 (2003)

SOMA T3A for Solving the 100-Digit Challenge

Quoc Bao Diep[1]([⊠]), Ivan Zelinka[1], Swagatam Das[2], and Roman Senkerik[3]

[1] Faculty of Electrical Engineering and Computer Science,
VSB - Technical University of Ostrava,
17. listopadu 2172/15, 708 00 Ostrava-Poruba, Czech Republic
diepquocbao@gmail.com, ivan.zelinka@vsb.cz
[2] Electronics and Communication Sciences Unit, Indian Statistical Institute,
203 Barrackpore Trunk Road, Kolkata, India
swagatam.das@isical.ac.in
[3] Faculty of Applied Informatics, Tomas Bata University in Zlin,
T. G. Masaryka 5555, 760 01 Zlin, Czech Republic
senkerik@utb.cz

Abstract. In this paper, we address 10 basic test functions of the 100-Digit Challenge of the SEMCCO 2019 & FANCCO 2019 Competition by using team-to-team adaptive seft-organizing migrating algorithm - SOMA T3A with many improvements in the Organization, Migration, and Update process, as well as the linear adaptive PRT and the cosine-based adaptive $Step$. The results obtained from the algorithm on the 100-Digit Challenge are very promising with 93 points in total.

Keywords: Self-organizing migrating algorithm · Optimization function · Swarm intelligence · SOMA T3A

1 Introduction

Self-organizing migrating algorithm (SOMA), a stochastic optimization algorithm, was inspired by cooperative and competitive behavior among intelligent creatures in the population such as birds and fish, to create new individuals that are candidate solutions to the problem. First published in 2000 [20], SOMA received much attention from researchers. There have been some publications, leading SOMA to not only having two initial strategies of AlltoOne and AlltoAll, but also making it more and more diverse and widely used. SOMA Pareto, the most recently published, is an example [8]. It applies the Pareto Principle in selecting individuals to be Migrants and individuals to be Leader. In addition, the adaptive control parameters have been proposed, significantly increasing the algorithm's performance, which has been proved by the well-known benchmark suite tests of the CEC'13, CEC'15, and CEC'17. In addition, other variant versions of SOMA such as M-SOMAQI [17], C-SOMGA [4], SOMAQI [18], mNM-SOMA [1], SOMGA [5], HSOMA [11], and CSOMA [16] have also confirmed their effectiveness.

© Springer Nature Switzerland AG 2020
A. Zamuda et al. (Eds.): SEMCCO 2019/FANCCO 2019, CCIS 1092, pp. 155–165, 2020.
https://doi.org/10.1007/978-3-030-37838-7_14

SOMA is widely used in many fields such as avoiding multiple dynamic obstacles and catching the moving target [2,9], solving large scale global optimization problems [12], complex network analysis [3], single and double layer spiral planar inductors optimization [13], and electromagnetic optimization [15]. However, problems become more and more complicated, current variants of SOMA are unable to solve these problems, or take a long time to achieve the required accuracy. The 100-Digit Challenge competition is an example [14].

The competition provides 10 hard functions, each participant is allowed to use an algorithm to solve (optimize) those functions with an accuracy of 10 digits each, and 10 points are awarded for each completely solved function. Different from previous CEC competitions, this year's competition highlights the accuracy rather than the time to solve. Therefore, the algorithm is not limited to function evaluations and is allowed to adjust 2 control parameters in the same way for all 10 functions. This competition is a great challenge for cutting-edge algorithms, see [14] for more detailed.

In this paper, we address 10 basic test functions of the 100-Digit Challenge of the SEMCCO 2019 & FANCCO 2019 Competition using team-to-team adaptive self-organizing migrating algorithm—SOMA T3A published in [6] with many improvements. The paper is structured into the following sections: Sect. 2 briefly describes the proposed algorithm, team-to-team adaptive SOMA, applied to solve the 100-Digit Challenge. Section 3 describes the experimental setting. The results are shown in Sect. 4. And finally, the paper is concluded with Sect. 5.

2   The Proposed Algorithm

2.1   Original Self-organizing Migrating Algorithm

This section briefly describes the Self-organizing migrating algorithm (SOMA), which is a premise for the proposed algorithm presented in the next subsection. SOMA is a stochastic optimization algorithm that mimics the competitive-cooperative intelligent behavior of creatures [10,19]. It works based on a population consisting of a given number of individuals that correspond to the candidate solution of the problem, and through many migration loops, these individuals interact with each other to search the best solution.

At the beginning of the algorithm, a population is randomly generated in the search range including a given number of individuals, as described in Eq. 1, and then the population will be evaluated by the given problem.

$$Pop = x_j^{(l)} + (x_j^{(h)} - x_j^{(l)})rand[0,1] \tag{1}$$

where:

- Pop: the population for the first migration loop,
- $x_j^{(l)}$: the lowest value of the boundary,
- $x_j^{(h)}$: the highest value of the boundary,
- j: from 1 to the population size,
- $rand[0,1]$: random number from 0 to 1.

An individual with the best value in the population is selected to be the Leader. All other individuals will move towards the Leader. After each finished movement, the best position on its path is selected to compare with its original position. If it is better, it will replace the original position. Otherwise, it will be ignored. Their movement process is described in Eq. 2.

$$x_n^{ML+1} = x_c^{ML} + (x_l^{ML} - x_c^{ML})\, t\, PRTVector, \tag{2}$$

where:

- x_n^{ML+1}: the new position of an individual in the next migration loop,
- x_c^{ML}: the position of this migrant in the current migration loop,
- x_l^{ML}: the position of the leader in the current migration loop,
- t: jumping step, from 0, by $Step$, to $PathLength$.

The $Step$ specifies the granularity of the migration process. The $PathLength$ determines how far individuals stop. By $PRTVector$ fixed and equal to 1, if $PathLength$ is greater than 1 (equal 3, for example) then this individual will stop behind the Leader and the distance from the initial position to this position is 3 times longer than the distance from it to the Leader. In fact, $PRTVector$ can be equal 1 or 0 depending on the PRT parameter, as in Eq. 3.

$$if\ rand < PRT;\ PRTVector = 1;\quad else,\ 0. \tag{3}$$

The PRT causes the individual instead of moving straight to the Leader, it will move in an N-k dimensional subspace. This helps the population preserve its diversity. The more PRT value goes to 1, the faster the algorithm will converge, but the higher it is to be trapped in the local minima and vice versa.

After all individuals complete the migration process, the next migration loop is started. Here, the best individual in the population is selected, and the moving process is continued until reaching the given stop conditions.

The mentioned strategy is named SOMA AllToOne. The strength of this strategy is its simplicity. For simple functions with small dimensions, SOMA AllToOne provides fast computing time and can find the global minima. But for complex functions and a larger number of dimensions, SOMA AllToOne can be trapped in the local minima, and no longer able to go beyond the local search subspace to find the global, as has been pointed out in [8].

Therefore, another method was proposed to address these weaknesses and apply it to solve the 100-Digit Challenge, named SOMA T3A [6].

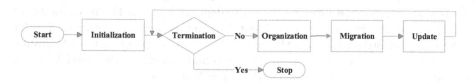

Fig. 1. The SOMA T3A flowchart.

2.2   SOMA T3A

This algorithm is divided into 4 processes: the Initialization, Organization, Migration, and Update process as shown in Fig. 1 [6].

Initialization Process. An initial population is randomly generated similar to SOMA AllToOne, and given in Eq. 1. After being initialized, these individuals are evaluated by the fitness function, in this case, 10 functions of the 100-Digit Challenge Competition.

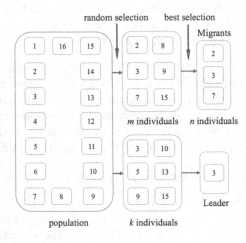

Fig. 2. The Organization process.

Organization Process. The mission of the Organization process is to determine which individuals will move, named Migrants, and which individual will be the possible Leader, that is, randomly select a small group containing m individuals in the population, and then select the best n out of m individuals to become Migrants. To select the Leader for each Migrant, the algorithm randomly select a small group containing k individuals in the population, and then select the best one from k, this individual is the Leader. The Organization process is described in Fig. 2.

There are two notes when choosing the Leader. First, individuals in the k group that includes the Leader may coincident with one of the n Migrants. In this case, the moving of this coincident Migrant is ignored. Secondly, Leader selection is performed for each Migrant. That is, corresponding to n Migrants, there are n times for selecting the k group and having n separated Leaders, instead of just choosing the only Leader for n Migrants.

This organization plays an important role in balancing the two phases of exploration and exploitation. If the values of m, n, and k are large, the algorithm will focus on the better individuals and tends to the exploitation phase. When their value is equal to the population size, the algorithm returning to its original version of SOMA AllToOne. Conversely, if their values are small, the algorithm will deviate to the exploration phase. There are no best values of m, n, and k for all optimization problems. Their values depend on the complexity and number of dimensions of functions, as well as the population size.

Migration Process. The Migration process describes how Migrants jump towards the Leader. In the SOMA AllToOne mentioned above, individuals jump in the N-k dimensional subspace instead of jumping straight to the Leader. The probability of each jump is determined by the PRT as in Eq. 3, and the step of each jump is a constant number. Individuals jump with fixed steps and fixed probability until reaching the $PathLength$. The restriction of two important parameters of the algorithm lies in the point: fixed value.

To propose a suitable PRT parameter, we consider the meaning of PRT. With the PRT parameter as close as 1, individuals have larger straight steps towards the Leader and fewer perpendicular steps in the N-k dimensional subspace. If PRT is equal to 1, individuals jump directly towards the Leader. This is suitable for use in the exploitation phase. On the other hand, the smaller the PRT, the larger the number of jumps in the N-k dimensional subspace, and the smaller the number of jumps straight to the Leader. This is suitable for use in the exploration phase. Since in the beginning, the algorithm should look for promising subspaces, and then, focus on that subspaces, so the PRT should start with a small value and gradually increases the value according to the migration loop. The adaptive PRT for the competition is described in Eq. 4.

$$PRT = 0.08 + 0.90 \ \frac{FEs}{MaxFEs} \tag{4}$$

where:

- FEs: the current number of function evaluation,
- $MaxFEs$: the maximun number of function evaluation.

The $Step$ determines the granularity of the jump. With a fixed number of steps, if the $Step$ is large, the algorithm will search on a wider subspace, and if the $Step$ is small, the algorithm will search for more detail in a narrower subspace. The adaptive $Step$ for the competition is described in Eq. 5.

$$Step = 0.02 + 0.005 \ cos(0.5\pi 10^{-7} FEs) \tag{5}$$

In the original version, note that the individual jumps until the $PathLength$ is reached. In this method, individuals jump until they reach the given number of jumps (named N_{jumps}), that means $PathLength = Step \times N_{jumps}$.

In the case of Migrants jumping out of the search range, a new random position will be generated in the search space to replace that position.

Update Process. The Update process is the process of updating new positions for each Migrant after it completes its jumping process. If this new position is better, it will replace the original position. In contrast, the Migrant remains in the initial position.

The whole process of the algorithm is described in Algorithm 1.

Algorithm 1. SOMA T3A

 1: Create the initial population
 2: Evaluate the initial population
 3: **while** stopping condition not reached **do**
 4:      Update *PRT* and *Step* values
 5:      Choose randomly m individuals from the population
 6:      Choose the best n Migrants out of m individuals
 7:      **for** $i = 1$ to n Migrants **do**
 8:          Choose randomly k individuals from the population
 9:          Choose the Leader from k individuals
10:          **if** the Migrant is not the Leader **then**
11:              The Migrant moves to the Leader
12:              Checking boundary
13:              Re-evaluate fitness function
14:              Updated the better position of the Migrant
15:          **end if**
16:      **end for**
17: **end while**
18: **return**

---

3   Experiment Setup

3.1   Implementation Environment

Matlab is used for programming and solving 10 basic test functions of the 100-Digit Challenge Competition of SEMCCO 2019 & FANCCO 2019 [14]. The test is performed on the Intel Core i7-6700 computer with 16 GB RAM, under the Windows 7 64-bit operating system, using Parallel Computing Toolbox of Matlab 2018a version (using 4 cores of CPU).

3.2   Functions

Table 1 lists the basic functions used in the 100-Digit Challenge. All test functions are scalable and they were designed to have the same global minimum value of 1.0 within the search range. The dimensions and search range are also in the two last columns in this table.

Table 1. The 100-Digit Challenge basic test functions.

No.	Functions	D	Search range
1	Storn's Chebyshev Polynomial Fitting Problem	9	$[-8192, 8192]$
2	Inverse Hilbert Matrix Problem	16	$[-16384, 16384]$
3	Lennard-Jones Minimum Energy Cluster	18	$[-4, 4]$
4	Rastrigin's Function	10	$[-100, 100]$
5	Griewangk's Function	10	$[-100, 100]$
6	Weierstrass Function	10	$[-100, 100]$
7	Modified Schwefel's Function	10	$[-100, 100]$
8	Expanded Schaffer's F6 Function	10	$[-100, 100]$
9	Happy Cat Function	10	$[-100, 100]$
10	Ackley Function	10	$[-100, 100]$

3.3 Scoring

SOMA T3A has been run 50 consecutive trials for each function, and the total number of correct digits in the 25 trials that had the best values has been counted. The average number of correct digits in 25 best trials is the score for that function. For example, the algorithm reaches equal or greater than 50% of the trials achieving 10 digits, then the score of that function is 10 points. The highest score in total is 100 points, refer to [14] for more details.

3.4 Termination Criterion

The algorithm will terminate when one of the two following criteria is met:

– achieving the 10-digit level accuracy,
– achieving the $MaxFEs$ ($MaxFEs = 10^9$ for 10 functions).

3.5 Control Parameters of SOMA T3A

Table 2 lists fixed parameters that were held constant value for 10 functions of the 100-Digit Challenge, containing the population size, the number of jumps, the number of individuals of the group n. The linear adaptive PRT and the cosine-based adaptive $Step$ are given in Eqs. 4 and 5.

The number of individuals of the group m and k are two tuned parameters. Their values, corresponding to each problem, are given in Table 3.

Different from the version [7] that also applied SOMA T3A to solve 10 basic test functions of the 100-Digit Challenge, two tuned parameters of the current version are the values of m and k instead of the $PopSize$ and the frequency of PRT. Besides the change of linear adaptive PRT and $MaxFEs$, this leads SOMA T3A to achieve better results than the older.

Table 2. The fixed parameters.

Parameters	Values
The population size ($PopSize$)	1500
The number of jumps (N_{jumps})	100
The number individuals of group n	4

Table 3. Two tuned parameters.

Function	1	2	3	4	5	6	7	8	9	10
m	50	50	50	50	50	50	50	5	20	50
k	100	50	100	100	100	100	80	5	30	100

4   Results

SOMA T3A has achieved 93 points in total in a run of 50 times for each function. Table 4 shows the detailed results, which the sequence number of functions are listed in the first column, the number of correct digits columns count the number of trials in a run of 50 times that the algorithm achieved from 1 to 10 correct digits respective to each function. The score for each function is the average number of correct digits of the best 25 runs that showed in the last column. The total score of all functions is presented in the last row.

In Table 4, SOMA T3A reached 50 over 50 runs achieving 10 correct digits on functions 1 to 6, and function 10. The score for those functions is 70 points (10 points for each function).

Table 4. Fifty runs for each function sorted by the number of correct digits

Function	Number of correct digits											Score
	0	1	2	3	4	5	6	7	8	9	10	
1	0	0	0	0	0	0	0	0	0	0	50	10
2	0	0	0	0	0	0	0	0	0	0	50	10
3	0	0	0	0	0	0	0	0	0	0	50	10
4	0	0	0	0	0	0	0	0	0	0	50	10
5	0	0	0	0	0	0	0	0	0	0	50	10
6	0	0	0	0	0	0	0	0	0	0	50	10
7	0	0	2	0	0	0	0	0	0	0	48	10
8	0	0	12	0	0	0	0	0	0	0	38	10
9	0	0	1	49	0	0	0	0	0	0	0	3
10	0	0	0	0	0	0	0	0	0	0	50	10
Total:												93

For functions 7 and 8, SOMA T3A has 2 and 12 times achieving 2 correct digits, has 48 and 38 times achieving 10 correct digits, respectively. The score for these functions is 20 points (10 for each of them).

Function 9 is the most difficult function for SOMA T3A when it only has 1 time achieving 2 correct digits and 49 times achieving 3 correct digits, there are no times in 50 runs that the algorithm achieves above 3 correct digits. The score for function 9 is 3 points.

In total, SOMA T3A achieved 93 points.

Figure 3 shows the mean function evaluations that the algorithm spent at each correct digit level. For functions 1, 2, 4, 5, 6, 7, and 10, the algorithm spent less than 2×10^8 FEs to achieve 10 correct digits. For functions 3 and 8, the algorithm spent nearly 8×10^8 FEs to achieve 10 correct digits. For function 9, the algorithm did not achieve 10 points and was not show in this figure.

From the data of Table 4 and Fig. 3, with function 9, the algorithm has been trapped in local minima and no longer able to go beyond the local subspace. For SOMA T3A within the current setting, function 3 and 8 are more complex than other functions, so the algorithm needs a longer time than the others to achieve 10 correct digits.

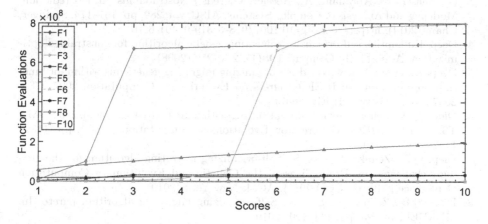

Fig. 3. The average number of FEs at each correct digit level (over the 25 best runs).

5 Conclusions

In this paper, we applied SOMA T3A to solve 10 problems of the 100-Digit Challenge of the SEMCCO 2019 & FANCCO 2019 Competition and reached 93 points in total. In this algorithm, SOMA was structured into the Initialization, Organization, Migration, and Update process. How to organize individuals in the population to become the Migrants and the Leader, as well as the linear adaptive PRT and the cosine-based adaptive $Step$ parameters, are the major improvements of this method beside an increment of $MaxFEs$ to 10^9. This helps the algorithm avoid being trapped in the local minima and reaching the good score in almost functions, except the function 9.

Acknowledgment. The following grants are acknowledged for the financial support provided for this research: Grant of SGS No. SP2019/137, VSB - Technical University of Ostrava. This work was also supported by the Ministry of Education, Youth and Sports of the Czech Republic within the National Sustainability Programme Project no. LO1303 (MSMT-7778/2014), further by the European Regional Development Fund under the Project CEBIA-Tech no. CZ.1.05/2.1.00/03.0089.

References

1. Agrawal, S., Singh, D.: Modified Nelder-Mead self organizing migrating algorithm for function optimization and its application. Appl. Soft Comput. **51**, 341–350 (2017)
2. Bao, D.Q., Zelinka, I.: Obstacle avoidance for Swarm robot based on self-organizing migrating algorithm. Procedia Comput. Sci. **150**, 425–432 (2019)
3. Davendra, D., Zelinka, I., Senkerik, R., Pluhacek, M.: Complex network analysis of discrete self-organising migrating algorithm. In: Zelinka, I., Suganthan, P.N., Chen, G., Snasel, V., Abraham, A., Rössler, O. (eds.) Nostradamus 2014: Prediction, Modeling and Analysis of Complex Systems. AISC, vol. 289, pp. 161–174. Springer, Cham (2014). https://doi.org/10.1007/978-3-319-07401-6_16
4. Deep, K.: Dipti: a self-organizing migrating genetic algorithm for constrained optimization. Appl. Math. Comput. **198**(1), 237–250 (2008)
5. Deep, K., et al.: A new hybrid self organizing migrating genetic algorithm for function optimization. In: IEEE Congress on Evolutionary Computation 2007, CEC 2007, pp. 2796–2803. IEEE (2007)
6. Diep, Q.B.: Self-organizing migrating algorithm team to team adaptive-SOMA T3A. In: 2019 IEEE Congress on Evolutionary Computation (CEC), pp. 1182–1187. IEEE (2019)
7. Diep, Q.B., Zelinka, I., Das, S.: Self-organizing migrating algorithm for the 100-digit challenge. In: Proceedings of the Genetic and Evolutionary Computation Conference 2019 (GECCO 2019). ACM, New York (2019)
8. Diep, Q.B., Zelinka, I., Das, S.: Self-organizing migrating algorithm pareto. In: MENDEL, vol. 25, pp. 111–120 (2019)
9. Diep, Q.B., Zelinka, I., Senkerik, R.: An algorithm for swarm robot to avoid multiple dynamic obstacles and to catch the moving target. In: Rutkowski, L., Scherer, R., Korytkowski, M., Pedrycz, W., Tadeusiewicz, R., Zurada, J.M. (eds.) ICAISC 2019. LNCS (LNAI), vol. 11509, pp. 666–675. Springer, Cham (2019). https://doi.org/10.1007/978-3-030-20915-5_59
10. Zelinka, I.: SOMA—self-organizing migrating algorithm. In: Davendra, D., Zelinka, I. (eds.) Self-Organizing Migrating Algorithm. SCI, vol. 626, pp. 3–49. Springer, Cham (2016). https://doi.org/10.1007/978-3-319-28161-2_1
11. Lin, Z., Juan Wang, L.: Hybrid self-organizing migrating algorithm based on estimation of distribution. In: 2014 International Conference on Mechatronics, Electronic, Industrial and Control Engineering (MEIC-14). Atlantis Press (2014)
12. Mohamed, A.W.: Solving large-scale global optimization problems using enhanced adaptive differential evolution algorithm. Complex Intell. Syst. **3**(4), 205–231 (2017)

13. Pospíšilík, M., Kouřil, L., Motýl, I., Adámek, M.: Single and double layer spiral planar inductors optimisation with the aid of self-organising migrating algorithm. In: Proceedings of the 11th WSEAS International Conference on Signal Processing, Computational Geometry and Artificial Vision, pp. 272–277. WSEAS Press (IT), Venice (2011)
14. Price, K.V., Awad, N.H., Ali, M.Z., Suganthan, P.N.: Problem definitions and evaluation criteria for the 100-digit challenge special session and competition on single objective numerical optimization. In: Technical report, Nanyang Technological University, Singapore, November 2018
15. dos Santos Coelho, L., Alotto, P.: Electromagnetic optimization using a cultural self-organizing migrating algorithm approach based on normative knowledge. IEEE Trans. Magn. **45**(3), 1446–1449 (2009)
16. dos Santos Coelho, L., Mariani, V.C.: An efficient cultural self-organizing migrating strategy for economic dispatch optimization with valve-point effect. Energy Convers. Manag. **51**(12), 2580–2587 (2010)
17. Singh, D., Agrawal, S.: Hybridization of self organizing migrating algorithm with quadratic approximation and non uniform mutation for function optimization. In: Das, K.N., Deep, K., Pant, M., Bansal, J.C., Nagar, A. (eds.) Proceedings of Fourth International Conference on Soft Computing for Problem Solving. AISC, vol. 335, pp. 373–387. Springer, New Delhi (2015). https://doi.org/10.1007/978-81-322-2217-0_32
18. Singh, D., Agrawal, S.: Self organizing migrating algorithm with quadratic interpolation for solving large scale global optimization problems. Appl. Soft Comput. **38**, 1040–1048 (2016)
19. Zelinka, I.: SOMA-self-organizing migrating algorithm. In: Onwubolu, G.C., Babu, B.V. (eds.) New Optimization Techniques in Engineering, pp. 167–217. Springer, Heidelberg (2004). https://doi.org/10.1007/978-3-540-39930-8_7
20. Zelinka, I., Jouni, L.: SOMA-self-organizing migrating algorithm mendel. In: 6th International Conference on Soft Computing, Brno, Czech Republic (2000)

Tracking the Exploration and Exploitation in Stochastic Population-Based Nature-Inspired Algorithms Using Recurrence Plots

Daniel Angus[1] and Iztok Fister Jr.[2]([✉])

[1] Digital Media Research Centre, Queensland University of Technology,
Brisbane, QLD 4059, Australia
[2] Faculty of Electrical Engineering and Computer Science, University of Maribor,
Koroška cesta 46, Maribor, Slovenia
`iztok.fister1@um.si`

Abstract. The success of every stochastic population-based nature-inspired algorithms is characterized through the dichotomy of exploration and exploitation. In general, exploration refers to the evaluation of points in previously untested areas of a search space, while exploitation refers to evaluation of points in close vicinity to previously visited points. How to balance both components properly during the evolutionary process is still considered as a topical problem in the evolutionary computation community. In this paper, we propose a recurrence plot visualization method for evaluating this process. Our analysis shows that recurrence plots are highly appropriate for revealing how particular algorithms balance exploration and exploitation.

Keywords: Exploration · Exploitation · Nature-inspired algorithms · Recurrence plot · Optimization

1 Introduction

Stochastic population-based nature-inspired algorithms are a kind of search algorithms that are considered as a powerful tool for coping with optimization problems in continuous, as well as discrete, domains. Most of them are inspired by the biological principles of behavior of various animals living in nature, while some of them are even inspired by physical phenomena. Each stochastic population-based nature-inspired algorithm consists of a population of individuals that undergo variation operators during the evolution process and generate a new subsequent population. Despite the popularity of this subject, a lot of different algorithms have been developed in the past decades. Nevertheless, characteristic examples that fit under this umbrella are: Artificial Bee Colony (ABC) algorithm [7], Bat Algorithm (BA) [16], Differential Evolution (DE) [12], Firefly Algorithm (FA) [15], Genetic Algorithm (GA) [5], Particle Swarm Optimization (PSO) [8].

© Springer Nature Switzerland AG 2020
A. Zamuda et al. (Eds.): SEMCCO 2019/FANCCO 2019, CCIS 1092, pp. 166–176, 2020.
https://doi.org/10.1007/978-3-030-37838-7_15

Each algorithm starts with a randomly generated initial population that updates over multiple generations/cycles using specific variation operators. For example, a GA uses three variation operators: selection, crossover and mutation; while the BA variation operator is guided by the physical phenomenon of echolocation observed in micro-bats. All of these operators influence the diversity of a population, and balance the exploration and exploitation components [1] and, most importantly, determine the overall quality of returned solutions. For this reason, it is important to have deep knowledge of the manner in which parameters of a particular algorithm impact on its search performance, which can help us understand what its weaknesses and advantages are during the evolutionary process. Additionally, such insights can help us to decide which algorithm is good for particular problems, as well as how to approach solving particular problems. We propose the use of recurrence plots to visualize graphically the evolutionary path of various nature-inspired algorithms, and reveal performance over time in a more informative manner than by simply tracking unitary measures such as the single best solution found. To the authors' knowledge, there is only one study [13] that used recurrence plots for study phase transitions in swarm optimization algorithms.

The main contributions of this paper are summarized as follows:

- to verify that there is a possibility to track the whole path of a stochastic population-based nature-inspired algorithm during the evolutionary process using recurrence plots,
- to investigate whether there is a possibility to observe changes between the exploration phase as well as the exploitation phase on recurrence plots,
- to study if there is a possibility to decide which algorithm is good for a particular problem based on the visualization of recurrence plots.

The structure of this paper is as follows: In Sect. 2 the stochastic population-based nature-inspired algorithms that are used in our study are outlined, while Sects. 3 and 4 present the methodology. The results of experiments are presented in Sect. 5. Section 6 concludes the paper, with remarks for future work.

2   Stochastic Population-Based Nature-Inspired Algorithms

The purpose of this section is to acquaint the reader with the population-based nature-inspired algorithms[1] that are being used in our experiments.

The Bat Algorithm is an example of Swarm Intelligence (SI) based algorithms [4]. BA is inspired by a physical phenomenon of micro-bats called echolocation. Differential Evolution is an evolutionary algorithm used widely in solving many combinatorial, continuous, as well as real-world problems. DE was proposed by Storn and Price in 1997 [12]. The Firefly Algorithm that was developed by Yang in 2008 is an SI-based algorithm inspired by the mating behavior of

---

[1] Sorted alphabetically.

fireflies. The phenomenon of fireflies is regarding the flashing lights that attract mating partners on the one hand, while, on the other, it serves as protection mechanism. Particle Swarm Optimization is also a member of SI-algorithms that was first presented in 1995 [8]. The inspirations of PSO lie in the social foraging behavior of some animals, such as the flocking behavior of birds.

3   Recurrence Quantification Analysis and Recurrence Plots

The recurrence plotting plot technique was initially invented as a technique to display and identify patterns from time series data, specifically data from high-dimensional dynamical systems [3]. The recurrence plot is a 2D plot where the horizontal and vertical axes represent time series data, and individual elements of the plot indicate times where the phase space trajectory of the system visits the same region of phase space.

While visual inspection of recurrence plots is useful for revealing the structure and dynamics of dynamical systems, Recurrence Quantification Analysis (RQA) extends this technique by specifying a set of metrics designed to capture specific features of recurrence plots [10,14]. In the 25 years following the original work of Eckmann et al. (1987), recurrence analysis has been applied across diverse areas including financial analysis, neural recordings, engineering, earth science and chemistry [9].

4   Methodology

In order to analyze recurrence plots for tracking the exploration and exploitation of population-based nature-inspired algorithms, we conducted a series of experiments. All experiments are based on the optimization of continuous benchmark functions [6] that are presented in Table 1.

Table 1. Benchmark functions used in our experiments.

f	Function name	Definition
$f1$	Sphere	$f(\mathbf{x}) = \sum_{i=1}^{D} x_i^2$
$f2$	Ackley	$f(\mathbf{x}) = -a \, \exp\left(-b\sqrt{\frac{1}{D}\sum_{i=1}^{D} x_i^2}\right) - \exp\left(\frac{1}{D}\sum_{i=1}^{D} cos(c \, x_i)\right) + a + \exp(1)$
$f3$	Griewank	$f(\mathbf{x}) = \sum_{i=1}^{D} \frac{x_i^2}{4000} - \prod_{i=1}^{D} \cos(\frac{x_i}{\sqrt{i}}) + 1$
$f4$	Rastrigin	$f(\mathbf{x}) = 10D + \sum_{i=1}^{D} \left(x_i^2 - 10\cos(2\pi x_i)\right)$

To generate a recurrence plot, a similarity measurement is required to compare any two points of the time series being plotted on the recurrence plot. In the case here, a single time point is the state of the population of the EC algorithm for a single generation, and, therefore, the similarity measurement is designed to measure the difference between two populations of solutions. The similarity measurement is based on the Euclidean distance between points in the population. The Euclidean distance metric is a relatively straightforward metric to calculate; given two solutions, p and q, and problem dimensionality of d, the Euclidean distance is calculated as:

$$\text{Euclidean distance}(p, q) = \sqrt{\sum_{i=1}^{d} (p_i - q_i)^2} \tag{1}$$

To obtain a population similarity score, Algorithm 1 is applied. According to the Algorithm 1, the algorithm returns a pairwise similarity score between any two solutions. Algorithm 1 sums the similarity of every solution between two populations. This sum is then divided by the population size. In line with this, Algorithm 2 presents how final time series are being generated. The algorithm iterates through all iterations, and calculates a similarity score for every two generations, i.e. the current generation and one next generation are taken into account. Finally, all points in the time series are normalized in order to get a similarity value between $[0, 1]$.

Algorithm 1. Population similarity score for populations $P1$ and $P2$

1: Score = 0
2: **for** $i = 1$ to NP_{P1} **do**
3:     **for** $j = 1$ to NP_{P2} **do**
4:         Score+ = Score $(P1_i, P2_j)$
5:     **end for**
6: **end for**
7: Score$/ = NP$

Algorithm 2. Building time series

1: TimeSeries = $\emptyset$;
2: **for** $i = 1$ to MAX_ITER **do**
3:     Point = $Calculate\_population\_similarity()$
4:     TimeSeries.$append$(Point)
5: **end for**
6: TimeSeries = $normalize(TimeSeries(0, 1))$

4.1   RQA Analysis and Generating a Recurrence Plot

Each population-based nature-inspired algorithm was run for 500 iterations, generating 20 new solutions per iteration. During the evolutionary cycle, we stored all solutions of each iteration. All included algorithms were run on 25 independent runs. Let us mention that the dimension of the problem was set to 30 for all algorithms on every benchmark function. Tables 2, 3, 4 and 5 present mean and std. values for each algorithm over the 25 runs. RQA was calculated using PyRQA software [11], while laminarity, divergence, trapping time and determinism measures were taken into account. For generating a recurrence plot, we chose 1 single run randomly from the pool of 25 runs. Recurrence plots were generated using the pyunicorn package [2].

RQA measures attempt to capture moments where a dynamical system under analysis is persisting in a single point in state space, drifting from or between different states, or randomly moving about a state space. RQA analysis is therefore of strong interest here due to its ability to capture aspects of convergent and non-convergent algorithmic behaviour. Laminarity is a measure of intermittent behaviour which will form vertical lines on a recurrence plot. Divergence is the inverse of the maximal diagonal line length which if low would indicate that an algorithm has converged or is moving along a cyclic trajectory through state space. Trapping time is the average length of vertical lines, which indicates the amount of time a system spends in a particular state, for an optimisation algorithm a high trapping time would indicate exploitation behaviour. Determinism is a percentage measure of how many recurrence points form diagonal lines. For determinism to be low a plot will contain mostly random noise (single recurrence dots), rather than longer diagonal lines which would indicate convergent behaviour, therefore low determinism indicates more exploration.

5   Discussion

For the algorithms tested we observe both qualitative and quantitatively different results from the recurrence analysis[2]. The BA algorithm produced some of the most interesting results, given that the algorithm seemed to converge within only

Table 2. RQA of BA.

Function	Measure	Laminarity	Divergence	Trapping time	Determinism
$f1$	Mean	0.9999	0.0026	366.1704	1.0000
	Std.	0.0001	0.0009	143.7695	0.0000
$f2$	Mean	0.9999	0.0021	474.5477	1.0000
	Std.	0.0000	0.0000	20.2955	0.0000
$f3$	Mean	0.9999	0.0023	450.7964	1.0000
	Std.	0.0001	0.0007	92.2660	0.0001
$f4$	Mean	0.9999	0.0021	468.0336	0.9998
	Std.	0.0004	0.0000	24.5921	0.0009

---

[2] Only selected figures are presented in this paper.

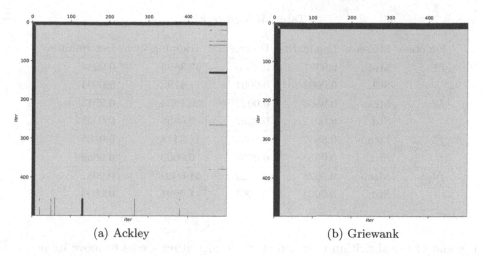

(a) Ackley   (b) Griewank

Fig. 1. Recurrence plots of BA on selected benchmark functions

a handful (10–30 iterations) leading to very large values for the RQA measures and almost complete visual recurrence (Fig. 1). At least in the configuration of the algorithm we used this would suggest that the BA algorithm is incredibly quick to converge and that care should be taken in ensuring that population diversity is maintained when in use.

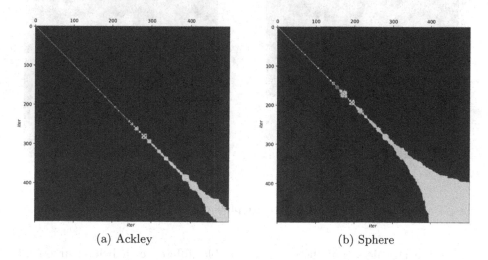

(a) Ackley   (b) Sphere

Fig. 2. Recurrence plots of FA on selected benchmark functions

The Firefly algorithm had the second highest values for Determinism, indicating that it too behaved in a highly exploitative fashion. The visual plots for this algorithm does reveal that this exploitation behaviour occurs mostly towards

Table 3. RQA of FA.

Function	Measure	Laminarity	Divergence	Trapping time	Determinism
$f1$	Mean	0.9927	0.0030	52.3893	0.9984
	Std.	0.0003	0.0001	1.4192	0.0004
$f2$	Mean	0.9566	0.0042	20.1938	0.9943
	Std.	0.0012	0.0002	0.6655	0.0021
$f3$	Mean	0.8857	0.0057	11.3448	0.9865
	Std.	0.0053	0.0004	0.6663	0.0059
$f4$	Mean	0.9926	0.0029	51.6059	0.9982
	Std.	0.0003	0.0001	1.0994	0.0003

the end of the algorithm run, and that the algorithm seems to move its population slowly through state space, highlighted also by the low divergence scores combined with high laminarity. Of all of the algorithms tested, FA tends to be the one algorithm that tends to transition the smoothest between exploratory and exploitative behaviours (Fig. 2).

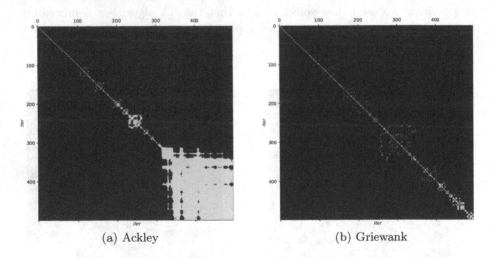

(a) Ackley   (b) Griewank

Fig. 3. Recurrence plots of DE 1/2

For the DE algorithm there was a notable difference in behaviour on F3 (Griewank), which can be seen both visually and quantitatively. On F3, DE seemed somewhat non-convergent, seen through the lower scores for laminarity, trapping time and determinism, and higher scores for divergence. the Griewank function quite notably contains a vast number of closely placed local minima, which could explain the algorithms lack of convergence (Figs. 3 and 4).

Table 4. RQA of DE.

Function	Measure	Laminarity	Divergence	Trapping time	Determinism
f1	Mean	0.9927	0.0028	39.8657	0.9916
	Std.	0.0048	0.0004	20.5044	0.0078
f2	Mean	0.9859	0.0052	72.6573	0.9947
	Std.	0.0140	0.0024	25.9202	0.0079
f3	Mean	0.6567	0.0453	4.6795	0.7223
	Std.	0.1937	0.0461	2.4264	0.1807
f4	Mean	0.8422	0.0349	20.4438	0.8379
	Std.	0.1005	0.0311	34.5553	0.1179

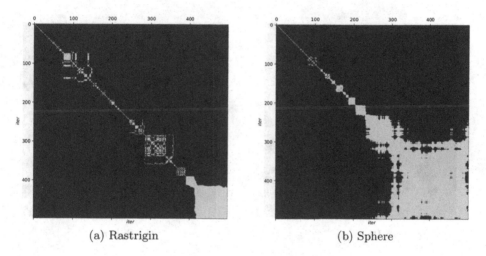

(a) Rastrigin   (b) Sphere

Fig. 4. Recurrence plots of DE 2/2

PSO had some of the most exploratory behaviour of all of the algorithms, with the highest divergence values, and lowest trapping time, determinism and laminarity. The recurrence plots for PSO reveal more information though, as one can quite clearly see how this algorithm seems to persist in distinct areas of the state space for tens of iterations. In the case of Rastrigin's function, it is clear that from iteration 200 the algorithm moves back and forth between areas of the state space creating what almost looks like a chess board pattern. Contrasted to FA, the results of PSO look more random and chaotic, rather than smoothly transitioning from one state to the next. In the case of the Sphere function, the PSO algorithm exhibits what could be considered a punctuated equilibrium effect moving from one point of state space, persisting for a time, then moving to another completely different section of state space (Fig. 5).

Table 5. RQA of PSO.

Function	Measure	Laminarity	Divergence	Trapping time	Determinism
$f1$	Mean	0.6419	0.0471	4.9304	0.6690
	Std.	0.1690	0.0446	1.9686	0.1788
$f2$	Mean	0.3901	0.1654	3.1237	0.4177
	Std.	0.2301	0.2094	1.0968	0.2347
$f3$	Mean	0.3309	0.1769	2.8711	0.3639
	Std.	0.1698	0.1992	0.8154	0.2046
$f4$	Mean	0.0377	0.8033	nan	0.0346
	Std.	0.0409	0.3003	nan	0.0647

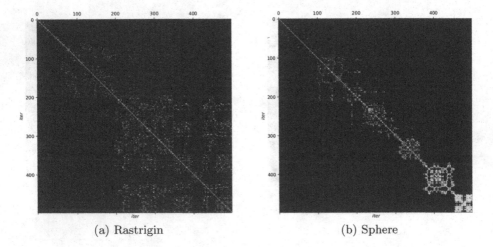

(a) Rastrigin   (b) Sphere

Fig. 5. Recurrence plots of PSO on selected benchmark functions

6   Conclusion

In this paper, we applied a well-known visualization technique, recurrence plotting, for tracking the exploration and exploitation of stochastic population-based nature-inspired algorithms. In addition, we also included the companion measures, Recurrence Quantification Analysis, which quantify distinct visual features of recurrence plots.

The resulting plots and RQA measures reveal much detail of the exploratory and exploitative behaviour of the algorithms under study. In the case of PSO, we could see much randomness, contained to a specific areas of search space, before moments where the algorithm jumped to a new area of search space to continue this behaviour anew. For FA, we noted a gradual shift from exploratory to exploitation as the algorithm progressed through its subsequent iterations. In DE disparity was seen in performance on different functions, indicating that recurrence plotting could help reveal disparity in performance within a single algorithmic class on different problems. And, in the case of BA we noted an almost instantaneous convergence behaviour, indicating an issue perhaps with the algorithms ability to trade off between exploration and exploitation over a single optimisation run.

These measures will benefit from more examination across more problem and algorithm classes, however through this modest study we have shown that these plots and measures can reveal much about the population dynamics of optimisation algorithms. The key difference between this and other unitary measures of algorithm performance, is that by taking full account of the population makeup, and change of this makeup over time, we can better ascertain an algorithms trajectory through state space. Future work could also examine the impact of different population similarity measurements on the resulting recurrence plots.

Acknowledgment. Iztok Fister Jr. acknowledge the financial support from the Slovenian Research Agency (Research Core Funding No. P2-0057).

References

1. Črepinšek, M., Liu, S.-H., Mernik, M.: Exploration and exploitation in evolutionary algorithms: a survey. ACM Comput. Surv. (CSUR) **45**(3), 35 (2013)
2. Donges, J.F., et al.: Unified functional network and nonlinear time series analysis for complex systems science: the Pyunicorn package. Chaos: Interdisc. J. Nonlinear Sci. **25**(11), 113101 (2015)
3. Eckmann, J.-P., Oliffson Kamphorst, S., Ruelle, D.: Recurrence plots of dynamical systems. EPL (Europhys. Lett.) **4**(9), 973 (1987)
4. Engelbrecht, A.P.: Computational Intelligence: An Introduction. Wiley, Hoboken (2007)
5. Goldberg, D.E.: Genetic Algorithms in Search, Optimization, and Machine Learning. Addison-Wesley Publishing Company, Boston (1989)
6. Jamil, M., Yang, X.-S.: A literature survey of benchmark functions for global optimisation problems. Int. J. Math. Model. Numer. Optim. **4**(2), 150–194 (2013)
7. Karaboga, D., Basturk, B.: A powerful and efficient algorithm for numerical function optimization: artificial bee colony (ABC) algorithm. J. Glob. Optim. **39**(3), 459–471 (2007)
8. Kennedy, J., Eberhart, R.: Particle swarm optimization. In: Proceedings of the IEEE International Conference on Neural Networks 1995, vol. 4, pp. 1942–1948. IEEE (1995)
9. Marwan, N., Romano, M.C., Thiel, M., Kurths, J.: Recurrence plots for the analysis of complex systems. Phys. Rep. **438**(5–6), 237–329 (2007)
10. Marwan, N., Wessel, N., Meyerfeldt, U., Schirdewan, A., Kurths, J.: Recurrence-plot-based measures of complexity and their application to heart-rate-variability data. Phys. Rev. E **66**(2), 026702 (2002)
11. Rawald, T., Sips, M., Marwan, N.: PyRQA-conducting recurrence quantification analysis on very long time series efficiently. Comput. Geosci. **104**, 101–108 (2017)
12. Storn, R., Price, K.: Differential evolution-a simple and efficient heuristic for global optimization over continuous spaces. J. Glob. Optim. **11**(4), 341–359 (1997)

13. Vantuch, T., Zelinka, I., Adamatzky, A., Marwan, N.: Phase transitions in swarm optimization algorithms. In: Stepney, S., Verlan, S. (eds.) UCNC 2018. LNCS, vol. 10867, pp. 204–216. Springer, Cham (2018). https://doi.org/10.1007/978-3-319-92435-9_15
14. Webber Jr., C.L., Zbilut, J.P.: Dynamical assessment of physiological systems and states using recurrence plot strategies. J. Appl. Physiol. **76**(2), 965–973 (1994)
15. Yang, X.-S.: Firefly algorithm, stochastic test functions and design optimisation. Int. J. Bio-Inspir. Comput. **2**(2), 78–84 (2010)
16. Yang, X.-S.: A new metaheuristic bat-inspired algorithm. In: González, J.R., Pelta, D.A., Cruz, C., Terrazas, G., Krasnogor, N. (eds.) Nature Inspired Cooperative Strategies for Optimization (NICSO 2010), pp. 65–74. Springer, Heidelberg (2010). https://doi.org/10.1007/978-3-642-12538-6_6

Insight into Adaptive Differential Evolution Variants with Unconventional Randomization Schemes

Roman Senkerik[1]([✉])[iD], Adam Viktorin[1][iD], Tomas Kadavy[1][iD],
Michal Pluhacek[1][iD], and Ivan Zelinka[2][iD]

[1] Faculty of Applied Informatics, Tomas Bata University in Zlin,
T. G. Masaryka 5555, 760 01 Zlin, Czech Republic
senkerik@utb.cz
[2] Faculty of Electrical Engineering and Computer Science, Technical University
of Ostrava, 17. listopadu 15, Ostrava, Czech Republic
ivan.zelinka@vsb.cz

Abstract. The focus of this work is the deeper insight into arising serious research questions connected with the growing popularity of combining meta-heuristic algorithms and chaotic sequences showing quasi-periodic patterns. This paper reports an analysis of population dynamics by linking three elements like distribution of the results, population diversity, and differences between strategies of Differential Evolution (DE). Experiments utilize two frequently studied self-adaptive DE versions, which are simpler jDE and SHADE, further an original DE variant for comparisons, and totally ten chaos-driven quasi-random schemes for the indices selection in the DE. All important performance characteristics and population diversity are recorded and analyzed for the CEC 2015 benchmark set in $30D$.

Keywords: Differential Evolution · Population diversity · Chaos-driven heuristics · CEC 2015 benchmark

1 Introduction

Ongoing research in metaheuristics algorithms is undoubtedly focused on hybridizations, extensive tuning, implementing learning strategies, and self-adaptive mechanism [25]. Outside this major research area, deterministic chaos with its properties like unique quasi-random sequencing and dynamics, quasi-stochasticity, self-similarity, fractal properties, and attractor density gained popularity as a simple technique for improving the metaheuristic algorithms performance.

The basic operation in the metaheuristic algorithms is randomness. Thus recent research in original/unconventional randomization techniques for metaheuristics mostly uses either directly scaled or normalized quasi-periodic sequences or a wide spectrum of different chaotic maps replacing the traditional uniform pseudo-random number generators (PRNGs). The importance of randomization within metaheuristics run has been profoundly investigated in several research papers, with the main focus

A. Zamuda et al. (Eds.): SEMCCO 2019/FANCCO 2019, CCIS 1092, pp. 177–188, 2020.
https://doi.org/10.1007/978-3-030-37838-7_16

either on describing different techniques for modification of the randomization process [2], or to influence of stochastic operations to the control parameters propagation [3].

The first study investigating the chaotic dynamic characteristics in swarm intelligence algorithms [4] and [5], that was later expanded in [6] was quickly followed by the general concept of chaos-driven genetic/evolutionary/swarm algorithms with embedded chaotic pseudo-random number generator (CPRNG) in [7]. Later, CPSO representing Particle Swarm Optimizer algorithm (PSO) with chaotic components [8], together with enhanced DE [9, 10], and inertia weight based PSO [11] laid the foundations of the popularity of the chaos embedded metaheuristics concept. Nowadays, it is very frequently used especially in real problem optimizations, where it is necessary to achieve a fine result quickly, mostly with simpler algorithms. Recently published research utilizes chaotic swarm algorithms [12–15] and also DE [16, 17]. Further, chaotic patterns in discrete dynamics of swarms have been investigated [18].

The next sections are focused on the motivation for this work, and differences with previous research papers, background of the DE algorithm, simple method for embedding CPRNG into DE, experiment setup, and detailed conclusions.

2   Related Work

The metaheuristic algorithm used in this research is Differential Evolution (DE) [1], specifically its frequently used simple, yet very powerful adaptive variant jDE and recent state of the art Success-History Based Adaptive Differential Evolution (SHADE).

The focus of this paper is the deeper insight into arising serious research questions connected with the analysis of population dynamics. The research reported here is linking three elements like distribution of the results, population diversity, and differences between strategies of DE versions. We have decided to track those features since the recent research in metaheuristics is focused on distance/diversity driven approaches [19–21] controlling search space exploration capabilities either through a distance between individuals, or through sustaining the population diversity at higher levels during the initial stages of the metaheuristic run.

However, the most important message of this research paper is that despite the still-growing popularity of the fusion of metaheuristic algorithms and unconventional randomization schemes (mostly with chaos), the majority of research papers do not explain, as to why those schemes have been used in the first place for enhancing the evolutionary operators like selection, mutation, crossover, or other processes (like communication in swarm).

This paper represents an incremental follow up of results and conclusions related to the population diversity analyses in chaotic DE published in [22, 23] and completes a recent work partially presented in [24]. The motivation, the difference from previously published work [24], and the originality of this paper are listed below:

- jDE and SHADE are investigated here in the dimensional settings of 30D. The abovementioned related works were mostly utilized simplest DE strategies and lower dimensional settings.

- The frequently used self-adaptive DE workflows enhanced by quasi-random chaotic sequencing as a randomization scheme is comprehensively reviewed here. This paper, extending the research in [24], could help navigate in differences between chaotic CPRNGs when embedded into self-adaptive DE schemes, and resulting performances as well as population dynamics.
- Here, advanced results analysis, which includes distribution plots and statistical rankings, is supporting conclusions.
- Direct comparisons between self-adaptive strategies and the original strategy of DE.
- The findings reported here can help build new approaches for improving exploration abilities and observing/analyzing possible causes of premature convergence through the population diversity, similarly, to the research reported in [25].

3 Differential Evolution

The generic DE [16] has several control parameters that remain static during the run and user setting dependent. The improved variants jDE and SHADE, which have been evolved from the generic DE algorithm, on the other hand, adapts the scaling factor F and crossover rate CR during the optimization (evolution process).

jDE is based on the propagation of two control parameters F_i and CR_i assigned to each i-th individual of the population. The basic idea of jDE lies in the "survival" of this parameter ensemble together with a successful solution. If an individual is transferred to the new generation, so is the parameter ensemble. If the newly generated solution is not successful; the control parameters pair disappears together with the lower quality solution. The above-mentioned pair of DE control parameters may be subject to random mutation based on user-defined probability.

State of the art variant SHADE utilizes a more advanced mutation strategy, self-adaptive mechanisms for control parameters adjustment based on historical memories and limited capacity archive for removed inferior solutions, that is based on elitism principle. The detailed description of essential operations in simple not-adaptive original DE is given in [1, 26], for the jDE, please see [27], and SHADE is detailed in [28].

4 Discrete Chaotic Systems as CPRNGs

The principle of applying the CPRNG is given by a simple exchange of the default algorithm/programming language PRNG with the deterministic chaotic system (preferably discrete one). This research is using the very same portfolio of chaotic maps as in [24]. With the definitions and internal parameters settings, as in [29], systems show expected chaotic dynamics and requested features. The example of discrete chaotic map definition is given in (1), representing the very popular and experimentally utilized Lozi map. The chaotic sequences of different length for the Lozi map is depicted in Fig. 1. These two plots show the presence of self-similarity within the chaotic sequence. Thus supporting the claim that the metaheuristics may be forced to neighborhood-based selection of individuals for evolutionary operators (or similarly, for neighborhood-based dynamics for spreading of information in swarms).

$$X_{n+1} = aX_n - Y_n^2$$
$$Y_{n+1} = bY_n + X_nY_n \tag{1}$$

The CPRNG workflow is as follows:

- Generating (by default algorithm/language PRNG) the starting position (X_0, Y_0) of the discrete chaotic map.
- Generating a chaotic sequence. The next iteration positions (X_{n+1}, Y_{n+1}) are obtained using their current positions (X_n, Y_n).
- Selection of particular sequence (x or y value, or combination of both) and re-normalization according to (2). When only the solo sequence is used (from two available), such a technique results in the folding of the chaotic attractor around the axis.

$$rndreal_n = \left| \frac{X_n}{maxval} \right| \tag{2}$$

Where the $rndreal_n$ is the re-normalized CPRNG value within the range of 0–1, here, we selected x-axis, and $maxval$ is the max. value from the whole chaotic series.

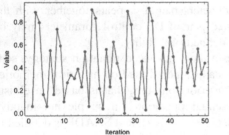 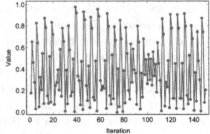

Fig. 1. Examples of two different CPRNG sequences, with significant patterns of self-similarity for Lozi map (50 iterations – left, and 150 iterations – right).

5   Experiment Setup

All executed instances (51 repetitions per instance) used the established CEC 15 test problems suite [30] with dimension D set to 30. The budget of FES was set to the value of 300 000 (10,000 × D), according to the general rules given in technical report for benchmark suite [30]. The performance features and population diversity were recorded for all executed variants of DE: generic DE, jDE and SHADE, further for nine chaotic versions of C_DE, C_jDE, and C_SHADE. The parameter setting for all algorithms is given in Table 1.

The identical set (as in [24]) of nine discrete dissipative chaotic maps were used here as the CPRNGs. The Population Diversity (PD) was also evaluated as in [24, 31].

Table 1. Parameter settings for DE variants

Parameter	DE	C_DE	jDE	C_jDE	SHADE	C_SHADE
NP (NP$_{init}$)	50	50	50	50	50	50
NP$_{min}$	N/A	N/A	N/A	N/A	N/A	N/A
max. gen.	6000	6000	6000	6000	6000	6000
MAXFES	300000	300000	300000	300000	300000	300000
F	0.5	0.5	0.5	0.5	N/A	N/A
CR	0.8	0.8	0.8	0.8	N/A	N/A
H	N/A	N/A	N/A	N/A	20	20
(C)PRNG	Java Linear congruential	9 different CPRNGs	Java Linear congruential	9 different CPRNGs	Java Linear congruential	9 different CPRNGs

6   Results

The statistical comparisons in comprehensive tables are not given here, as this was not the main aim of this paper. Instead, the rankings of the algorithms are presented in Fig. 2, evaluated based on the *Friedman test with Nemenyi post hoc test*. Further, distribution plots for 51 runs are depicted in Figs. 3 and 4, where the left column shows original DE, middle one jDE and right column contents results for SHADE of CEC15; functions from upper to bottom, *f1–f7* in Fig. 3 and *f8–f15* in Fig. 4.

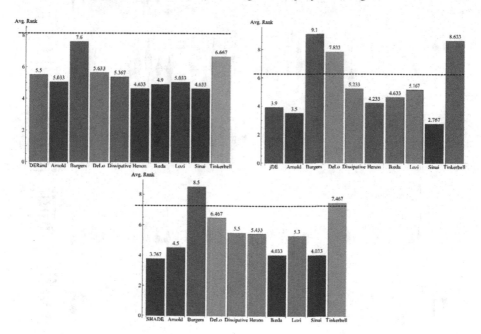

Fig. 2. Ranking of all algorithms (DE – upper left, jDE – upper right, and SHADE – below center), 51 runs, 15 functions of the CEC2015 test suite in 30*D*. The dashed line represents the Nemenyi Critical Distance.

Figures 5 and 6 show the heat maps for the population diversity recorded and analyzed for the first 500 generations and selected subset of CEC 15 functions (where the differences between versions are the most visible). Such an interval was selected since we can assume that the initial stage of the evolutionary process may be critically sensitive to the keeping of the population's diversity at higher values securing the search space exploration. The detailed results discussion is presented in the conclusion section.

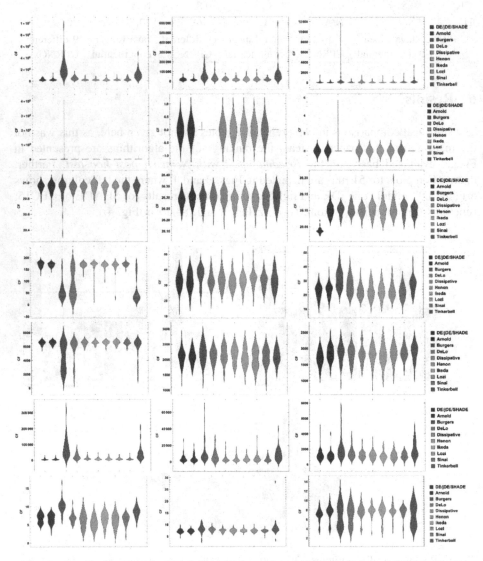

Fig. 3. DistributionPlots for DE (left), jDE (center) and SHADE (right) of CEC15 (*f1–f7*)

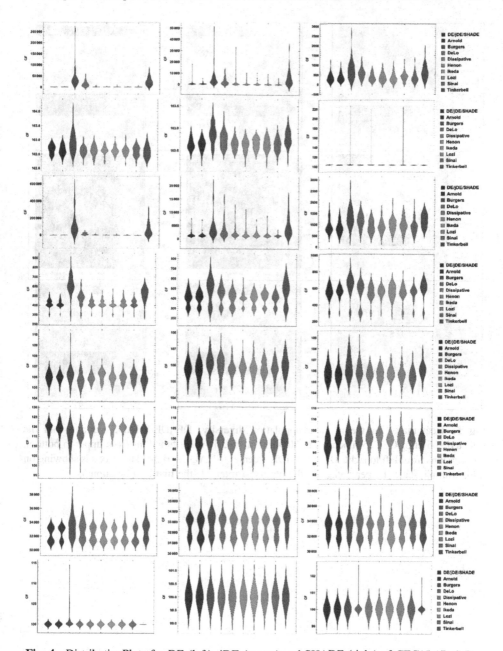

Fig. 4. DistributionPlots for DE (left), jDE (center) and SHADE (right) of CEC15 (*f*8–*f*15)

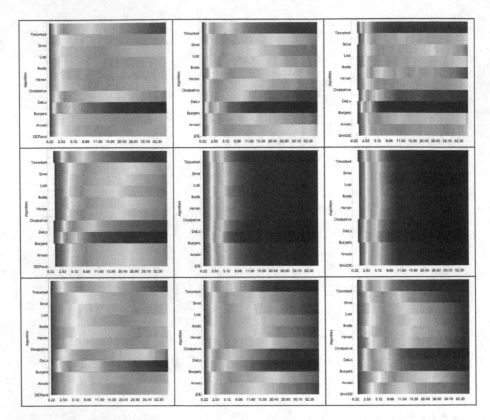

Fig. 5. Heat maps for the average population diversity, DE/jDE/SHADE variants and the selected subset of CEC15 functions in 30*D* (*f*3, *f*4, *f*5), 51 runs; functions from upper to bottom, DE versions from left to right: DE – left, jDE – center, SHADE – right. The *x*-axis is showing on a logarithmic scale the percentage of the observed interval (the first 500 iterations).

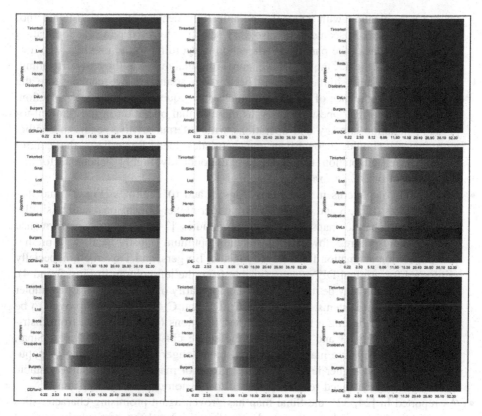

Fig. 6. Heat maps for the average population diversity, DE/jDE/SHADE variants and the selected subset of CEC15 functions in 30D (f7, f13, f14), 51 runs; functions from upper to bottom, DE versions from left to right: DE – left, jDE – center, SHADE – right. The x-axis is showing on a logarithmic scale the percentage of the observed interval (the first 500 iterations).

7 Conclusion

This paper completes a recent work [22, 24, 32] with more DE versions, higher dimensional settings ($D = 30$) and different detailed processing of results. Research and experiments with population dynamics and randomization schemes have been addressed as of high importance and advantageous. Therefore, graphical heat maps, distribution plots, and rankings of all used algorithms represent here a deeper investigation into the population dynamics and selection of individuals inside self-adaptive DE variants, which are jDE and SHADE algorithms. Specifically, under the influence of different types of quasi-random (chaotic) sequences.

The findings below may reveal alternative ways of developing new ideas and more effective metaheuristics. The findings are suggested based on the CEC 2015 testbed. Under other settings, of course, these might turn out different.

- The rankings of algorithms depicted in Fig. 2 indicate that the original fully stochastic mutation scheme "Rand/1/" implemented in jDE has proven to be the favorable choice for the hybridization with original randomization mechanisms. The original (not adaptive) DE variant seems to be also very sensitive to the driving randomization scheme, but it lacks the parameter adaptation. According to the rankings, the not-chaotic jDE was the 3rd, the C_jDE version with Sinai map was the best performing.

- In SHADE variant (similarly to any best-driven strategies) we can observe the suppression of influence of quasi-random/chaotic sequencing for indices selection. The attraction to the *"p*best" solution as the essential operation in the "current-to-pbest/1" mutation strategy and operations with external archive seems to conflict with general simple chaos-driven metaheuristics idea. Regardless, the control parameter self-adaptation techniques are undoubtedly highly beneficial for the improvement of DE performance.

- Correlations of patterns in heat maps (Figs. 5 and 6), rankings (Fig. 2), and distribution plots (Figs. 3 and 4) reveals here under higher-dimensional setting similar clustering of chaotic maps as in previous studies [24, 32]: worse performance showing rapid progress towards test function optimum (mostly local) with premature population stagnation for Delayed logistic, Burgers, and Tinkerbell. Especially reinforced for original DE strategy (visible from the left column in Figs. 3 and 4). This phenomenon may be the answer to the popularity of chaotic systems in simpler algorithms/swarm systems. Most of the C_jDE and C_SHADE have proven to be very effective in finding a minimum of test functions (as can be seen from distribution plots - Figs. 3 and 4). Further, the stable cluster of Arnold, Lozi, Sinai, Henon, and Dissipative maps has secured for a longer time the keeping of population diversity at higher levels. Thus this group of maps showed relatively stable and statistically balanced results compared to the non-chaotic DE versions.

- As reported in [24, 32], the Sinai map has proven again promising characteristics for some test cases, where the population diversity has been particularly restored during the run of DE (and the exploration phase could be restarted).

- SHADE shows a rapid decrease in diversity (rapid convergence) and much less sensitivity to randomization than jDE and the original DE.

- Chaotic/quasi-periodic sequencing influences the population diversity and may lead to the creation of the sub-populations (or some quasi-periodic neighborhood individual selection schemes [33]).

The results presented here support the development of advanced randomization mechanisms (multi-chaotic generators [34]) or ensemble randomization systems [35], providing the combined/selective population diversity for effective controlling of exploration/exploitation during the run of metaheuristic algorithm.

The popular self-adaptive DE versions have been tested here, and we can observe, that besides the adaptive/learning-based control, other approaches can be effectively used to achieve better/desired DE performance.

Acknowledgement. Authors RS, AV, TK and MP acknowledge the support of project No. LO1303 (MSMT-7778/2014) by the Ministry of Education, Youth and Sports of the Czech Republic within the National Sustainability Programme, further the project CEBIA-Tech no. CZ.1.05/2.1.00/03.0089 under the European Regional Development Fund. Authors AV and TK also acknowledge the Internal Grant Agency of Tomas Bata University under the project

No. IGA/CebiaTech/2019/002. This work is also based upon support by COST Action CA15140 (ImAppNIO), and the resources of A.I.Lab at the Faculty of Applied Informatics, TBU in Zlin (ailab.fai.utb.cz). Finally, Author Ivan Zelinka acknowledges the support of grant SGS 2019/137, VSB-Technical University of Ostrava.

References

1. Price, K.V., Storn, R.M., Lampinen, J.A.: Differential Evolution: A Practical Approach to Global Optimization, Berlin. Springer, Heidelberg (2005). https://doi.org/10.1007/3-540-31306-0
2. Weber, M., Neri, F., Tirronen, V.: A study on scale factor in distributed differential evolution. Inf. Sci. **181**(12), 2488–2511 (2011)
3. Zamuda, A., Brest, J.: Self-adaptive control parameters' randomization frequency and propagations in differential evolution. Swarm Evol. Comput. **25**, 72–99 (2015)
4. Meng, H.J., Zheng, P., Mei, G.H., Xie, Z.: Particle swarm optimization algorithm based on chaotic series. Control Decis. **21**(3), 263 (2006)
5. Liu, H., Abraham, A., Clerc, M.: Chaotic dynamic characteristics in swarm intelligence. Appl. Soft Comput. **7**(3), 1019–1026 (2007)
6. Liu, H., Abraham, A.: Chaos and swarm intelligence. In: Kocarev, L., Galias, Z., Lian, S. (eds.) Intelligent Computing Based on Chaos, pp. 197–212. Springer, Heidelberg (2009). https://doi.org/10.1007/978-3-540-95972-4_9
7. Caponetto, R., Fortuna, L., Fazzino, S., Xibilia, M.G.: Chaotic sequences to improve the performance of evolutionary algorithms. IEEE Trans. Evol. Comput. **7**(3), 289–304 (2003)
8. Coelho, L., Mariani, V.C.: A novel chaotic particle swarm optimization approach using Hénon map and implicit filtering local search for economic load dispatch. Chaos Solitons Fractals **39**(2), 510–518 (2009)
9. Davendra, D., Zelinka, I., Senkerik, R.: Chaos driven evolutionary algorithms for the task of PID control. Comput. Math Appl. **60**(4), 1088–1104 (2010)
10. Ozer, A.B.: CIDE: chaotically initialized differential evolution. Expert Syst. Appl. **37**(6), 4632–4641 (2010)
11. Pluhacek, M., Senkerik, R., Davendra, D.: Chaos particle swarm optimization with Eensemble of chaotic systems. Swarm Evol. Comput. **25**, 29–35 (2015)
12. Wang, G.G., Deb, S., Gandomi, A.H., Zhang, Z., Alavi, A.H.: Chaotic cuckoo search. Soft. Comput. **20**(9), 3349–3362 (2016)
13. Fister Jr., I., Perc, M., Kamal, S.M., Fister, I.: A review of chaos-based firefly algorithms: perspectives and research challenges. Appl. Math. Comput. **252**, 155–165 (2015)
14. Alatas, B.: Chaotic bee colony algorithms for global numerical optimization. Expert Syst. Appl. **37**(8), 5682–5687 (2010)
15. Metlicka, M., Davendra, D.: Chaos driven discrete artificial bee algorithm for location and assignment optimisation problems. Swarm Evol. Comput. **25**, 15–28 (2015)
16. Zhang, J., Lin, S., Qiu, W.: A modified chaotic differential evolution algorithm for short-term optimal hydrothermal scheduling. Int. J. Electr. Power Energy Syst. **65**, 159–168 (2015)
17. Mokhtari, H., Salmasnia, A.: A Monte Carlo simulation based chaotic differential evolution algorithm for scheduling a stochastic parallel processor system. Expert Syst. Appl. **42**(20), 7132–7147 (2015)
18. Das, S.: Chaotic patterns in the discrete-time dynamics of social foraging swarms with attractant–repellent profiles: an analysis. Nonlinear Dyn. **82**(3), 1399–1417 (2015)

19. Viktorin, A., Senkerik, R., Pluhacek, M., Kadavy, T., Zamuda, A.: Distance based parameter adaptation for success-history based differential evolution. Swarm Evol. Comput. **50**, 100462 (2018)
20. Sudholt, D.: The benefits of population diversity in evolutionary algorithms: a survey of rigorous runtime analyses. arXiv preprint arXiv:1801.10087 (2018)
21. Corus, D., Oliveto, P.S.: Standard steady state genetic algorithms can hillclimb faster than mutation-only evolutionary algorithms. IEEE Trans. Evol. Comput. **22**(5), 720–732 (2018)
22. Senkerik, R., Viktorin, A., Pluhacek, M., Kadavy, T., Zelinka, I.: How unconventional chaotic pseudo-random generators influence population diversity in differential evolution. In: Rutkowski, L., Scherer, R., Korytkowski, M., Pedrycz, W., Tadeusiewicz, R., Zurada, J.M. (eds.) ICAISC 2018. LNCS (LNAI), vol. 10841, pp. 524–535. Springer, Cham (2018). https://doi.org/10.1007/978-3-319-91253-0_49
23. Senkerik, R., Viktorin, A., Pluhacek, M., Kadavy, T., Oplatkova, Z.K.: Differential evolution and chaotic series. In: 2018 25th International Conference on Systems, Signals and Image Processing (IWSSIP), pp. 1–5. IEEE, June 2018
24. Senkerik, R., et al.: Population diversity analysis in adaptive differential evolution variants with unconventional randomization schemes. In: Rutkowski, L., Scherer, R., Korytkowski, M., Pedrycz, W., Tadeusiewicz, R., Zurada, J.M. (eds.) ICAISC 2019. LNCS (LNAI), vol. 11508, pp. 506–518. Springer, Cham (2019). https://doi.org/10.1007/978-3-030-20912-4_46
25. Karafotias, G., Hoogendoorn, M., Eiben, Á.E.: Parameter control in evolutionary algorithms: trends and challenges. IEEE Trans. Evol. Comput. **19**(2), 167–187 (2014)
26. Das, S., Mullick, S.S., Suganthan, P.: Recent advances in differential evolution – an updated survey. Swarm Evol. Comput. **27**, 1–30 (2016)
27. Brest, J., Greiner, S., Bosković, B., Mernik, M., Zumer, V.: Self-adapting control parameters in differential evolution: a comparative study on numerical benchmark problems. IEEE Trans. Evol. Comput. **10**(6), 646–657 (2006)
28. Tanabe, R., Fukunaga, A.S.: Improving the search performance of SHADE using linear population size reduction. In: 2014 IEEE Congress on Evolutionary Computation (CEC), pp. 1658–1665. IEEE (2014)
29. Sprott, J.C.: Chaos and Time-Series Analysis. Oxford University Press, Oxford (2003)
30. Chen, Q., Liu, B., Zhang, Q., Liang, J.J., Suganthan, P.N., Qu, B.Y.: Problem definition and evaluation criteria for CEC 2015 special session and competition on bound constrained single-objective computationally expensive numerical optimization. Technical report, Computational Intelligence Laboratory, Zhengzhou University, China and Nanyang Technological University, Singapore (2014)
31. Poláková, R., Tvrdík, J., Bujok, P., Matoušek, R.: Population-size adaptation through diversity-control mechanism for differential evolution. In: MENDEL, 22th International Conference on Soft Computing, pp. 49–56 (2016)
32. Senkerik, R., Viktorin, A., Pluhacek, M., Kadavy, T.: On the population diversity for the chaotic differential evolution. In: 2018 IEEE Congress on Evolutionary Computation (CEC), pp. 1–8. IEEE, July 2018
33. Das, S., Abraham, A., Chakraborty, U., Konar, A.: Differential evolution using a neighborhood-based mutation operator. IEEE Trans. Evol. Comput. **13**(3), 526–553 (2009)
34. Viktorin, A., Pluhacek, M., Senkerik, R.: Success-history based adaptive differential evolution algorithm with multi-chaotic framework for parent selection performance on CEC2014 benchmark set. In: 2016 IEEE Congress on Evolutionary Computation (CEC), pp. 4797–4803. IEEE, July 2016
35. Wu, G., Mallipeddi, R., Suganthan, P.N.: Ensemble strategies for population-based optimization algorithms–a survey. Swarm Evol. Comput. **44**, 695–711 (2019)

Virtual Measurement of the Backlash Gap in Industrial Manipulators

Eliana Giovannitti[1]([✉])(iD), Giovanni Squillero[1](iD), and Alberto Tonda[2,3](iD)

[1] Politecnico di Torino, Turin, Italy
{eliana.giovannitti,squillero}@polito.it
[2] Université Paris-Saclay, Saclay, France
[3] UMR 782 Inria, Thiverval-Grignon, France
alberto.tonda@inra.fr

Abstract. Industrial manipulators are robots used to replace humans in dangerous or repetitive tasks. Also, these devices are often used for applications where high precision and accuracy is required. The increase of backlash caused by wear, that is, the increase of the amount by which teeth space exceeds the thickness of gear teeth, might be a significant problem, that could lead to impaired performances or even abrupt failures. However, maintenance is difficult to schedule because backlash cannot be directly measured and its effects only appear in closed loops. This paper proposes a novel technique, based on an Evolutionary Algorithm, to estimate the increase of backlash in a robot joint transmission. The peculiarity of this method is that it only requires measurements from the motor encoder. Experimental evaluation on a real-world test case demonstrates the effectiveness of the approach.

Keywords: Evolutionary computation · Backlash · Robotic joint transmission · Shaft variable stiffness

1 Introduction

In an industrial context it is of paramount importance to guarantee correct and continuous operation of machinery, as in complex production lines—consisting of hundreds of devices—any abrupt stop may lead to significant economic losses. To this regard, *industrial manipulators*, the robots used to replace humans in dangerous or repetitive tasks, are particularly critical: an extremely high precision and accuracy is usually required, while gears are inevitably subjected to mechanical deterioration.

The *backlash* is the rotational arc clearance between a pair of mating gear teeth, that is, the amount by which a tooth space exceeds the thickness of a gear tooth engaged in mesh. While a small amount of backlash is intentionally designed to ensure smooth movements, the increase of the gear play due to wear may cause important nonlinearities and eventually limit the performance of speed controllers, possibly causing even a permanent damage to the apparatuses.

© Springer Nature Switzerland AG 2020
A. Zamuda et al. (Eds.): SEMCCO 2019/FANCCO 2019, CCIS 1092, pp. 189–200, 2020.
https://doi.org/10.1007/978-3-030-37838-7_17

With industrial manipulators, an estimate of the backlash gap is useful to foresee possible criticalities and perform the maintenance before a malfunctioning or a breakdown. Unfortunately in most robots the backlash appears in closed loop and typically there is access only to the final motor speed [3].

This paper addresses the problem of estimating the increase of the backlash in a robot joint transmission relying only on measures on the motor encoder and estimates based on models and Evolutionary algorithms.

The paper is organized as follows. In Sect. 2 the problem of the backlash in a mechanical transmission and the evolutionary algorithm on which its estimation is based are introduced. Section 3 describes the mechanical system under examination, its Matlab/Simulink model that allowed the analysis of the backlash phenomenon and its effect on the motor speed signal. Finally the set-up for the genetic algorithm used for the backlash estimation is exposed. In Sect. 5 the results of the proposed method, applied to the case of two speed signals measured on the real mechanical system, are presented.

2   Background

2.1   Backlash

Gears are used to transmit torque from the motor to the load. In an ideal gear system the mating gear teeth are always in contact, perfectly transmitting movement from the motor to the load. In the presence of backlash the contact between two paired teeth is interrupted for a small angle and then it is re-established. This can cause impacts and vibrations on the moving parts and a lower positioning accuracy for the robot. Many backlash mathematical models are available in literature, the classical *dead zone model* and *hysteresis model* [3,4] are among the most used ones.

The model we used in this work is presented in [5]. It is a modification of the *dead zone model* and has been integrated into the Simulink model of the mechanical system under test. The system is represented with the typical linear dynamic model used for a mechanical transmission: a two-mass system with an elastic coupling and backlash. The first mass represents the motor, with the moment of inertia J_m, that is coupled to the load, the second mass with moment of inertia J_l, by a shaft. The shaft is considered mass free and is modeled with a torsional stiffness spring K_s and a damping D_S. The backlash gap is δ (Table 1 and Fig. 1).

When the mating gears are in contact the motor is connected to the load, the load torque τ_l is proportional to the angle difference $\Delta\theta$ and to the speed difference $\Delta\omega$, see Eq. (1). When the gear travels the backlash gap the motor loses contact with the load and the load torque becomes zero, see Eq. (2).

Defining

$$\Delta\theta = \frac{1}{N}\theta_m - \theta_l$$

$$\Delta\omega = \frac{1}{N}\omega_m - \omega_l$$

Table 1. System parameters.

Symbol	Description	Units
θ_m, θ_l	Motor/Load angular position	rad
ω_m, ω_l	Motor/Load angular velocity	rad
τ_m, τ_l	Motor/Load torque	Nm
J_m, J_l	Motor/Load inertia	Kg m^2
K_s	Shaft stiffness	Nm/rad
D_s	Shaft damping coefficient	Nm s/rad
δ	Backlash angle	rad
N	Gear ratio	–

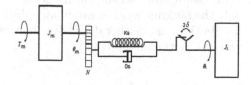

Fig. 1. Two mass system with elastic coupling and backlash.

The interconnecting torque τ_l is

$$\tau_l = K_s \Delta\theta + D_s \Delta\omega \tag{1}$$

and in presence of backlash Eq. (1) becomes

$$\tau_l = \begin{cases} K_s(\Delta\theta - \delta \cdot sign(\Delta\theta)) + D_s\Delta\omega & |\Delta\theta| > \delta \\ 0 & |\Delta\theta| \leq \delta \end{cases} \tag{2}$$

2.2 Covariance Matrix Adaptation Evolution Strategy

Covariance Matrix Adaptation Evolution Strategy (CMA-ES) is a stochastic optimization technique belonging to the family of Evolutionary Algorithms (EAs). EAs [1] are loosely inspired by the neo-Darwinian paradigm of natural selection, and are able to efficiently explore large and irregular search spaces. As they are stochastic, there is no guarantee that they will find any global optimum, but in several complex real-world problems they proved able to deliver solutions of high quality in a reasonable amount of time, and nowadays are applied to problems in which traditional optimization techniques fail [7].

EAs also belong to the family of *local search algorithms*: they starts by generating random candidate solutions to the problem, and their capability to solve the problem at hand is measured by a *fitness function*. In successive iterations, good candidate solutions are more likely to be selected to *reproduce*, generating new candidate solutions that are similar to the originals. The process is repeated until a solution of satisfying quality is found, or until a user-specified stop condition is reached.

The field of evolutionary computation can be further divided into categories of algorithms, such as Genetic Algorithms (GAs) or Genetic Programming (GP) – a reminiscent of their origin. Among such algorithms, *Evolution Strategies* (ES) [6,8] soon emerged as a quite powerful tool for optimizing problems with real-valued variables. CMA-ES [2] is the most effective extension of the original algorithm available nowadays. In CMA-ES, the adaptation of the covariance matrix amounts to learning a second order model of the underlying objective function similar to the approximation of the inverse Hessian matrix in the Quasi-Newton method in classical optimization. In contrast to most classical methods, fewer assumptions on the nature of the underlying objective function are made. Only the ranking between candidate solutions is exploited for learning the sample distribution and neither derivatives nor even the function values themselves are required by the method.

3   Proposed Approach

The proposed approach is composed of five independent steps:

- The backlash phenomenon is evaluated from a theoretical perspective. A simulation model is built to assess its effect on the measured speed.
- The disturbance pattern is represented as an analytic expression with a set of parameters.
- The disturbance pattern at increasing backlash values is generated by simulation, and a relationship between the relevant parameters and the actual backlash value is determined.
- The real speed signal is recorded. The parameters of the disturbance pattern are evaluated by fitting the theoretical disturbance on the measured data by an EA.
- The estimated parameters are eventually used to assess the backlash value affecting the real system.

A common approach to the backlash analysis of a mechanical transmission is the use of an output (i.e. on the load side) encoder or a torque sensor. These devices allow a direct measure of the quantities of interest. The present method instead only relies on the encoder provided on motor side which is the standard equipment for an industrial robot. The amount of backlash is estimated by using the motor speed signal. By defining proper working and stress conditions it is possible to detect the presence of a disturbance signal superimposed on the speed signal. The disturbance has a known aspect and can be related to the backlash phenomenon. The disturbance waveform has been identified, isolated and validated by a test campaign on the test bench that reproduces the real transmission of an industrial manipulator joint.

4   Experimental Setup

A Matlab/Simulink model of the entire system was created, containing a mathematical model of the backlash phenomenon. Through this model it was investigated how the disturbance evolves as the backlash increases. Furthermore, the relationship that links the intensity of the disturbance to the value of the backlash gap was identified. Taking advantage of this relationship, it was possible to estimate the backlash angle amplitude starting from the motor speed signal. The estimate is performed using an evolutionary algorithm (CMA-ES).

4.1   Test Bench

The system under test is a typical rotary joint of an industrial manipulator. It is composed of a motor, a transmission belt and a reducer with a backlash value exceeding the defined acceptable limits. A load consisting of a cast iron mass is connected to the system. The only accessible measures on the system are the motor position, provided by an encoder connected to the motor, and the current absorbed by the motor itself (Fig. 2). The system has no additional sensors after the transmission to obtain a direct measurement of the backlash. The disturbance appears as an undesirable oscillation on the speed signal measured by the encoder.

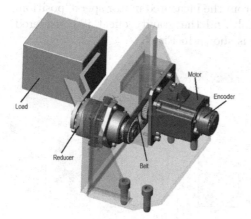

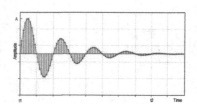

Fig. 2. Test Bench

Fig. 3. Model of the disturbance signal induced by the backlash

The disturbance oscillations have the typical appearance of a percussive phenomenon in an elastic system with damping. Similarly to what happens when a hammer hits the string of a piano, the impact generates oscillations with an amplitude that decays exponentially with time (Fig. 3).

For this reason it was decided to attribute the mathematical model described by the following formula to the disturbance:

$$d_b(t) = \begin{cases} 0 & t < t_1 \\ A\,e^{-(t-t_1)\tau} \sin \omega(t) & t_1 \leq t \leq t_2 \\ 0 & t > t_2 \end{cases} \qquad (3)$$

where
t_1 = the stating time of the oscillation
A = the maximum amplitude of the oscillation
τ = a damping factor
t_2 = the ending time of the disturbance.

Tests were conducted by running the motor at a constant speed. Under these conditions the system is in a steady state in which the effects of disturbances such as static frictions and inertial phenomena are not present. In these circumstances the backlash phenomenon is highlighted due to small impacts caused by the action of gravity.

4.2   Matlab/Simulink Model

Once the working conditions and the input signal have been defined and the shape of the disturbance has been identified, a Matlab/Simulink model of the entire system was developed. The model used for the system is the one presented in Sect. 2.1 and is composed by: a motor with an encoder, a transmission, a load and a control loop with linear feedback from the measured motor speed/position. The transmission is affected by backlash and the gravity effect is considered through the load dynamics. The model is shown in Fig. 4.

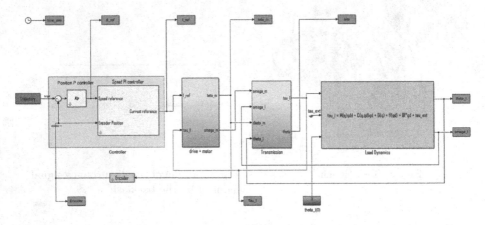

Fig. 4. Matlab/Simulink Model of the system

By leveraging the simulation flexibility it was possible to analyze different backlash conditions and to understand how the disturbance changes as the backlash increases. This helped to identify the relation between the intensity of the disturbance and the value of the backlash.

Many different backlash models have been proposed, the one used in this work belongs to the *deadzone type* and represents the backlash in terms of variable stiffness [5]. Outside the backlash zone τ_l is proportional to the angle difference between motor and load multiplied by the shaft stiffness. When the gear tooth travels the backlash zone the load disengages from the driving motor and the torque τ_l becomes zero.

The effect expressed by the Eq. (2) can be also achieved through a variable shaft stiffness that becomes zero in the backlash zone:

$$K_{BL}(\Delta\theta, \delta) = \frac{K_s}{\pi}[\pi + \arctan(\alpha(\Delta\theta - \delta)) - \arctan(\alpha(\Delta\theta + \delta))]$$

$K_{BL}(\Delta\theta, \delta)$ is depicted in Fig. 5.

The model employs the *arctan* function to avoid abrupt discontinuities. Acting on the α factor, a positive constant, it is possible to change the *arctan* slope. The load torque then becomes:

$$\tau_l = [\Delta\theta - \delta \cdot sign(\Delta\theta) + \frac{D_s}{K_s}\Delta\omega] \cdot K_{BL}(\Delta\theta, \delta)$$

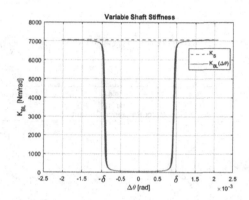

Fig. 5. Variable shaft stiffness. Stiffness value is K_S outside the deadzone, $|\Delta\theta| > \delta$, and then goes to zero when $|\Delta\theta| \le \delta$

The model parameters used for the simulator setup were given by mechanical data and identification experiments.

To test the Simulink model we compared the simulation output with the measurements on the real system. Signals comparison is showed in Fig. 6. The signals in light color are the position, the speed and the current of the motor measured on the test bench. The signals in the darker color are the position, the speed and the current of the motor obtained by the Simulink model. The figure shows that the simulator is able to correctly reproduce the real behaviour of the system with backlash.

Using simulation and varying the value of the backlash within an interval $\delta = [\delta_{min}, \delta_{max}]$ it was noted that the amplitude A of the oscillation of the disturbance signal (Eq. (3)) is directly linked to the backlash amount δ. It was therefore possible to find a relationship between A and δ which allows to estimate the value of the backlash once the disturbance amplitude has been identified. Given the mechanical properties of our system a reasonable choice for the delta interval was $[0.0001, 0.0040]$ *radians*.

The simulation results and the relation $A \mapsto \delta$ are showed in Figs. 7 and 8. The regression analysis results showed that the relation $\delta(A)$ is well described by a cubic polynomial:

$$\delta(A) = 10^{-4}(-0.000009\,A^3 + 0.003122\,A^2 + 0.138764\,A - 0.138809) \quad (4)$$

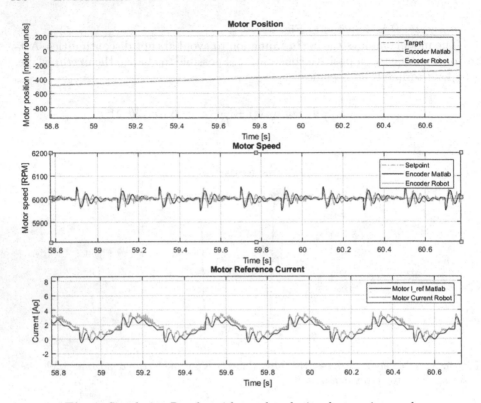

Fig. 6. Simulation Results with test bench signals superimposed.

4.3 Backlash Identification

EA was used to recognize the backlash pattern in the motor speed signal. The identification relies on the minimization of the error between a measured signal and a generalization of the model defined in Eq. (3). The *CMA-ES algorithm*, described in Sect. 2.2, was used for this activity. This tool is available on *GitHub*[1] in a Python implementation.

The main idea is to minimize the RMS error between the real signal $v(t)$ and the model we developed for the signal with backlash, Eq. (6). We generalized the $d_b(t)$ model, see Eq. (3), by defining a new function translated by a time offset t_0:

$$g(t) = d_b(t - t_0) \tag{5}$$

Since gravity acts on the system as a torque on the motor that varies sinusoidally, the backlash causes a pulsed periodic disturbance. The time period corresponds to a 2π rad load rotation and contains two pulses having opposite signs. This effect is due to the link hitting the gear and the gear hitting the link

---

[1] https://github.com/CMA-ES/pycma.

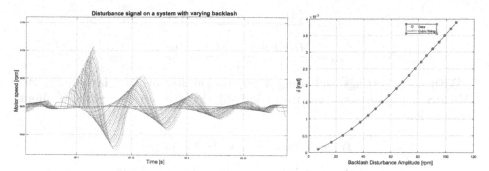

Fig. 7. Disturbance appearance at increasing backlash gap.

Fig. 8. Disturbance oscillation amplitude, A, and backlash gap value, δ, regression.

at the beginning of respectively the descending and the ascending phases of the load movement. This led to the definition of the function

$$f(t) = g(t) - g(t + T)$$

Identification was performed on a sequence of 4 disturbance repetition in order to obtain mean values for the parameter identification. So the final model was

$$h(t) = v_t + \sum_{i=1}^{4} f(t - i \cdot T_w)$$

where T_w is the time interval that corresponds to a full load rotation and v_t is the target motor speed.

The function relies on 7 unknown parameters and can be expressed as

$$h(t, A, t_0, \tau, \omega, v_t, t_1, t_2, T, T_w) = v_t + \sum_{i=1}^{4} f(t, A, t_0 + i \cdot T_w, \tau, \omega, t_1, t_2, T) \quad (6)$$

The function to be minimized is then

$$RMSE = \sqrt{\frac{\Sigma_{i=1}^{N} \left(v(t) - h(t, A, t_0, \tau, \omega, v_t, t_1, t_2, T, T_w) \right)^2}{N}}$$

For a fast algorithm convergence, it is critical to properly define the initial conditions for the parameters. We set up a procedure to compute the initial values based only on the measured signal $v(t)$ and a limited a priori system knowledge. The procedure starts by computing the variability range of each parameter (see Table 2) and then estimates the starting value for the mean x_0 and the variance σ_0 of each parameter as

$$x_0 = \frac{Max\_value + Min\_value}{2}$$

$$\sigma_0 = \frac{Max\_value - Min\_value}{4}$$

With these settings and using a population size of 3500 individuals the *CMA_ES algorithm* converges to a good solution (i.e. error = 0.14) in about 100 iterations.

Table 2. Parameters variability range

Symbol	Min_value	Max_value	Units
A	$-\frac{\max v(t) - \min v(t)}{2}$	$\frac{\max v(t) - \min v(t)}{2}$	rpm
t_0	$\min t$	$\max t$	s
τ	5	30	–
ω	2π	$2\pi \cdot 40$	rad/s
v_t	$\min v(t)$	$\max v(t)$	rpm
t_1	0	0.204	s
t_2	0	$\frac{0.204}{2}$	s
T	$\min t$	$\max t$	s
T_w	$\min t$	$\max t$	s

NOTE: t is the time vector of the measured signal v(t)

5 Experimental Results

Starting from two different datasets $v_1(t)$ and $v_2(t)$, acquired on the test bench and corresponding to two different and progressive situations of wear, it was possible to detect the increase in backlash gap through the use of CMA-ES. The algorithm was able to recognize the known disturbance pattern within the speed signal and to estimate the value of its parameters.

The two different identifications returned two increasing values for the oscillation amplitude A:

$$A_1 = 30.9033 \, [rpm] \, , \quad A_2 = 39.2197 \, [rpm]$$

and, through the Formula (4), the corresponding backlash values:

$$\delta_1 = 6.8539e - 04 \, [rad] \, , \quad \delta_2 = 9.5394e - 04 \, [rad].$$

The results are shown in Fig. 9 and in Fig. 10.

Figure 9 illustrates how, for both datasets, CMA_ES was able to obtain an average value of the model parameters, allowing to correctly approximate the starting signal, even if affected by a high noise. Moreover, Fig. 10 shows that the algorithm is able to detect changes in parameter A with sufficient sensitivity to characterize changes in the backlash gap over time. The two datasets, in fact, were obtained from the test bench at a distance of about 4 months during a continuous operation cycle of the device.

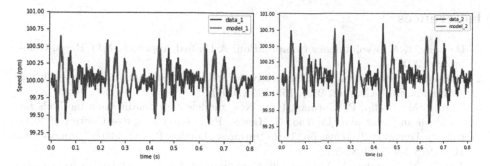

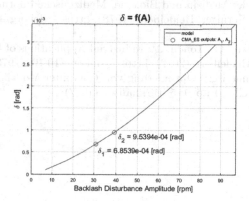

Fig. 9. CMA_ES identification results for dataset_1 (on the left side) and dataset_2 (on the right side). The signals acquired on the test bench are plotted in blue color; red plots are used for the model reconstruction relying on the parameters identified with CMA_ES. (Color figure online)

Fig. 10. Final backlash evaluation.

6   Conclusions and Future Works

A method for estimating the backlash in a mechanical transmission of a robot joint has been presented. The strategy used was explained and the results demonstrate the effectiveness of the proposed method. The result of the estimation can be used to implement strategies to compensate for the disturbance deriving from the backlash or to diagnose the operation of the robotic manipulator. The estimation of the parameters was performed using a state-of-the-art stochastic optimization technique, the CMA-ES.

The next step will be to embed a small device on each manipulator, able to record data and either transmit them to a centralized server or to process them in-situ. The required computational power is limited, as the CMA-ES may fit the disturbance parameters starting from a previously found solution.

References

1. De Jong, K.A.: Evolutionary Computation: A Unified Approach. MIT Press, Cambridge (2006)
2. Hansen, N., Ostermeier, A.: Completely derandomized self-adaptation in evolution strategies. Evol. Comput. **9**(2), 159–195 (2001)
3. Nordin, M., Bodin, P., Gutman, P.O.: New models and identification methods for backlash and gear play. In: Tao, G., Lewis, F.L. (eds.) Adaptive Control of Nonsmooth Dynamic Systems, pp. 1–30. Springer, London (2001). https://doi.org/10.1007/978-1-4471-3687-3_1
4. Nordin, M., Gutman, P.O.: Controlling mechanical systems with backlash-a survey. Automatica **38**, 1633–1649 (2002)
5. Papageorgiou, D., Blanke, M., Niemann, H.H., Richter, J.H.: Backlash estimation for industrial drive-train systems. IFAC-PapersOnLine **50**(1), 3281–3286 (2017)
6. Rechenberg, I.: Evolutionsstrategien. In: Schneider, B., Ranft, U. (eds.) Simulationsmethoden in der Medizin und Biologie. Medizinische Informatik und Statistik, vol. 8, pp. 83–114. Springer, Heidelberg (1978). https://doi.org/10.1007/978-3-642-81283-5_8
7. Sanchez, E., Squillero, G., Tonda, A.: Industrial Applications of Evolutionary Algorithms. Springer, Heidelberg (2012). https://doi.org/10.1007/978-3-642-27467-1
8. Schwefel, H.P.: Numerische Optimierung von Computer-Modellen mittels der Evolutionsstrategie, Teil 1, Kap. 1–5. Birkhäuser (1977)

Hybrid Elephant Herding Optimization Approach for Cloud Computing Load Scheduling

Ivana Strumberger⊙, Eva Tuba⊙, Nebojsa Bacanin⊙, and Milan Tuba$^{(\boxtimes)}$⊙

Singidunum University, 11000 Belgrade, Serbia
{istrumberger,nbacanin}@singidunum.ac.rs, {etuba,tuba}@ieee.org

Abstract. Cloud computing is rather important distributing computing paradigm and in general refers to the common pool of configurable resources that is accessed on-demand. Resources are dynamically scalable and metered with the basic aim to provide reliable and quality services to the end-users. Load scheduling has a great impact on the overall performance of the cloud system, and at the same time it is one of the most challenging problems in this domain. In this paper, we propose implementation of the hybridized elephant herding optimization applied to load scheduling problem in cloud computing. The algorithm is using CloudSim framework, and comparison with different metaheuristics, adapted and tested under same experimental conditions, for this type of problem was performed. Moreover, we compared proposed hybridized elephant herding optimization with its original version in order to evaluate its improvements in performance over the original version. Obtained empirical results prove the robustness and quality of approach that we propose in this paper.

Keywords: Cloud computing · Swarm intelligence · Load scheduling

1 Introduction

Distributed computing, as one of the domains in computer science, studies distributed systems, which consist out of components that are implemented on different devices in the computer network. In most cases, these components communicate and coordinate their processes by using a message passing mechanism. Due to reliability, high availability, scalability, efficiency and lower costs of computing resources, distributed computing has in recent decade gained significant attention from the industry, as well as from the academic community.

One of the most important distributed computing paradigm is cloud computing. In general, the term cloud computing refers to the access in on-demand manner to a collective pool of customizable resources that are dynamically scalable and metered towards the basic objective of providing to the end-users quality and reliable services. Basics of cloud computing had been devised in the

© Springer Nature Switzerland AG 2020
A. Zamuda et al. (Eds.): SEMCCO 2019/FANCCO 2019, CCIS 1092, pp. 201–212, 2020.
https://doi.org/10.1007/978-3-030-37838-7_18

mainframe computing era, when "dump terminals" were utilizing CPU, memory and other resources from the distant mainframe system. Using a concept of cloud computing, assets in terms of software and hardware are delivered via the computer network (in most cases the Internet) to the consumers.

The technology of virtualization and hyper-converge infrastructure (HCL) enable cloud computing concept. One instance of many available definitions from the modern literature states that the virtualization represents an abstraction of software and hardware that breaks traditional architecture of computer system. This emerging technology has a significant role in latest computer platforms [1]. Virtualization technology provides the means of decoupling operating system and applications, that are executing on top of the physical computer hardware, from the hardware itself. By using virtualization, as state-of-the-art technology, multiple virtual machines (or virtual instances) may run in a pseudo-parallel manner on the same physical host in its own isolated environment.

The virtual machine manager (VMM), or hypervisor, is a component which is responsible for creating, terminating and managing virtual instances, while at the same time also performs some other functions in the virtual environment. In the real, production environemnts, two hypervisors' categories exist: type 1 (level 1) and type 2 (level 2). In the modern literature, for type 1 and type 2 hypervisors, are usually used terms bare metal and hosted hypervisors, respectively. The basic difference between the type 1 and type 2 hypervisors is that the type 1 runs just above the physical hardware infrastructure, while the latter is executed within the environment of host operating system. Due to its efficiency compared to the level 2 hypervisor, most enterprise environments utilize type 1. Some of the most representative examples of bare metal VMMs are Microsoft Hyper-v, Xen/Citrix XenServer and VMware vSphere/ESXi, while the hosted hypervisors such as VMWare Workstation, Parallels and Oracle Virtual Box, are commonly used in home, as well as in small enterprise environments.

Second technology, that enables cloud computing paradigm, refers to the software-defined computer platform, that performs virtualization of traditional physical hardware based components. As a minimum requirements, the HCL encompasses SDS (software-defined storage), hypervisor and SDN (software-defined networks). Only by utilizing vritualization technology and HCL, cloud services provider (CSP) is able to deliver quality and reliable services to the consumers in a cost-effective way via computer network.

1.1   Cloud Computing Paradigm and Related Work

Many definitions of cloud computing can be found by surveying modern scientific sources. According to one definition, the cloud computing is pool of virtual hardware and software that enable and transport (deliver) various services to the consumers [2]. One of other available definitions describes cloud computing system as distributed elastic system where the information, software, storage space, and other resources are scattered throughout the network and can be approached and shared by many consumers at the same time from distant cloud locations [3]. Due to the fact that cloud provides variety of delivered assets, the concept of cloud computing as a model was widely used in the industry and academics.

Scalability and availability of assets and tasks, fault tolerance, load balancing, services execution on demand, and resource interoperability represent only some of the most essential cloud computing attributes. Cloud resources can be utilized with the proper management activities of virtual resources on the cloud.

When describing the cloud computing concept, it should be noted that many categorizations of cloud computing exist, among which two are the most important. First categorization takes into account the models of service criteria and distinguishes cloud services between infrastructure as a service (IaaS), platform as a service (PaaS) and software as a service (SaaS). SaaS refers to the accessed via standard interfaces software, such is Web-based interface, which is configured in a hosted service form. By utilizing PaaS, that encompasses operating systems and platforms, end-users may implement and configure requested applications on the cloud platform. Finally, the IaaS refers to the cloud service model, that delivers to the clients infrastructure components, such as storage space, networking and CPU time. Second categorization divides cloud computing services into public, private, community and hybrid, by utilizing the delivery models criteria.

The most basic goal of the CSPs is to obtain promised quality of service (QoS) to the clients with a constraint that cloud computing ought to be economically efficient way of employing computing resources for both parties, end-users and the CSP. Computational and cost efficiency problems are the main issues with the load or tasks scheduling and balancing which represent significant challenges in cloud computing environment [6].

By using scheduling algorithms, near optimal, or optimal distribution (allocation) of available capabilities between requested assignments can be obtained within satisfying time range to obtain desired QoS [7]. The goal of scheduling is to generate a plan which defines on which available resources and when every tasks will be deployed for execution.

In cloud computing environment, the task scheduling or load balancing problem refers to the optimal allocation of submitted end-users' tasks to the finite set of available virtual instances. Since task scheduling is NP-hard challenge, algorithms that can accomplish optimal feasible solution in a polynomial time do not exist. In such cases, the best approach is to employ metaheuristics that are able to generate reasonable solutions in a satisfying time range.

Swarm intelligence, that simulates collective behavior of organisms from the nature, falls into the category of nature inspired metaheuristics. As population-based, iterative optimization approach, swarm intelligence has shown great potential in tackling NP hard benchmarks [8,9], as well as practical optimization problems [10,11]. Some of the representative examples of swarm algorithms include: artificial bee colony (ABC) [12], bat and cuckoo search algorithms [16], firefly algorithm (FA) [13–15]. Based on the current state in the literature, swarm intelligence was successfully used for solving challenges in cloud computing in the previously conducted researches [17–19, 28].

1.2   Paper Goal and Organization

The essential goal of proposed paper is to present and use implementation of the hybridized elephant herding optimization algorithm (EHO) for cloud computing task scheduling. EHO, as relatively novel swarm intelligence approach, was devised by Wang in 2015 [20]. We enhanced the original version of the EHO metaheuristics by adopting crossover operator from the genetic algorithm (GA) and devised genetic EHO (GEHO) metaheuristics.

Proposed metaheuristics was tested in the CloudSim framework environment that is widely used as a research tool. We utilized model of task scheduling and CloudSim parameters like in the research shown in [3], and compared the results with different methods that were presented in [3], as well as with the original EHO algorithm.

This paper is structured into five sections. After Introduction, Sect. 2 presents task and load scheduling model that was employed for testing purposes of the proposed metaheuristics. Details of the orignal EHO, as well as the hybridized EHO (GEHO) were given in Sect. 3. Empirical results and comparative test with other outstanding algorithms are given in Sect. 4. Finally, in Sect. 5 we present conclusions from our research, along with the guidelines for future work.

2   Load Scheduling Optimization Model

When overlooking the topological aspect of cloud environment, every data center in the cloud has definite number physical hosts or servers with heterogeneous configurations. These hosts have many typical features such as identifier of a host (ID), processing elements number, performance of processing, usable memory, and many other characteristics and features. A server (or host) runs numerous virtual instances (VMs) pseudo-simultaneously, providing diverse heterogeneous applications. The request for cloud resources from end-user first comes to the main balancer of the load, which schedules and maps cloud assets to the appropriate virtual machines. Each request is appointed to one available VM in the pool. Virtual machine is assigned to the end-users' requests and afterwards upon the execution, it becomes accessible for another task.

Since there are numerous cloud end-users' resource requirements that can occur at exact same time slot, with various input tasks, the load balancing mechanism is imperative for this purpose. The process of these requirements is as follows: the n input assignments $T_0, T_1, ..., T_{n-1}$ are arranged in the assignments queue within the cloud, which are afterwards sent to the VM manager. The manager possess the data about the servers availability, and active VMs, that are to be scheduled for these tasks. It is essential for the VM manager to verify system resources available for tasks to be executed, whether they are sufficient or not. In the case when the tasks can be executed on accessible active virtual machines, the manager transfers those cloud assignments to the load and task scheduler. If that's not the case, the VM manager then makes new VMs on physical servers that have sufficient resources.

If NVM VMs are being employed by the system to schedule NC tasks (cloudlets), the overall number of potential distributions of tasks to virtual machines may be expressed as NVM^{NC}, and that is a NP hard problem. When tackling this type of problem, each individual from the population is determined as the d-dimensional vector, where the number of dimensions depends on the number of used virtual instances. Objective of the proposed metaheuristics is achieving higher exploitation of virtual machines by the tasks, that results in decisive load scheduling. In the model that is used in this paper, the total computation cost and the total transfer time have been advised as objectives. The objective function is defined as the computation cost along with the transfer cost of the tasks and virtual machines. Objective function for individual i at the moment t is presented in the Eqs. (1)–(6) [3], where:

$$costExec(M_j) = \sum_k w_{k,j} \forall M(k) = j \tag{1}$$

$$costTransfer(M_j) = \sum_{k1 \in T} \sum_{k2 \in T} d_{M(k1),M(k2)} \tag{2}$$

$$\cdot e_{k1,k2} \forall M(k1) = j \text{ and } M(k2) \neq j \tag{3}$$

$$costTotal(M_j) = costExec(M_j) + costTransfer(M_j) \tag{4}$$

$$totalMaxCost(M) = \max(costTotal(M_j)) \forall j \in P \tag{5}$$

$$\min(totalMaxCost(M) \forall M) \tag{6}$$

The notations $costTransfer(M_j)$ and $costExec(M_j)$ present transfer and execution costs for the j-th assignment, respectively. The total cost $costTotal(M_j)$, which is the addition of the transfer and execution cost, is being determined for each task, and afterwards, the total maximum cost $totalMaxCost(M)$ is being minimized [3].

3 Original and Hybridized Elephant Herding Optimization Metaheuristics

The group behavior of elephants in herds inspired the creation of elephant herding optimization (EHO) algorithm, that was devised in 2015 by Wang et al. for tackling bound-constrained tasks [20]. There are diverse implementations of this metaheuristics in many domains, such as practical problems as static drone placement problem [21], node localization in WSNs [22], multilevel image threshold [23] as well as implementation and testing on standard benchmark problems [24] etc. The essential communication and social coexistence in the herd depends on the leadership of the matriarch, where structural bond depends on matriarch's influence. When male calves grow, they depart but important thing to note is that they still communicate with the herd, expanding the bond and social intelligence. This occurring defines the general-purpose heuristic search, which comprises these two phenomenons in the herds change.

These environments can be divided into the first environment (updating operator), where elephants are coexisting within the leadership of the matriarch, and the second (separating operator), in which full grown male elephants leave the herd.

Every individual j in each clan, denoted as c_i, is updated by the present location and the fittest member of c_i by using an updating operator. The first phase is followed by improvement of the population discrepancy by employing the separating operator in the next generations of the algorithm's execution.

The whole population consists of the n clans. The updating operator is denoted by changing each solution j in the each clan c_i by the impact of the fittest solution in c_i which can be seen in Eq. 7:

$$x_{new,c_i,j} = x_{c_i,j} + \alpha \times (x_{best,c_i} - x_{c_i,j}) \times r, \tag{7}$$

where $x_{new,c_i,j}$ is the new location for the individual j in the clan c_i, and x_{best,c_i} is the fittest solution of the c_i at the time of calculation, where $x_{c_i,j}$ denotes the previous position of the solution j that belongs to the clan ci. Parameter $\alpha \in [0,1]$ represents a scale indicator that designates the authority of the best solution in c_i on $x_{c_i,j}$, and $r \in [0,1]$ is a random number. To update the best solution in clans the following equation is used:

$$x_{new,c_i,j} = \beta \times x_{center,c_i}, \tag{8}$$

where $\beta \in [0,1]$ presents the impact of the x_{center,c_i} on the new generated solution.

The parameter D denotes the search space overall dimension, following is the calculation of the clan c_i center, $x_{center,c_i,d}$ for problem with $d - th$ dimensions:

$$x_{center,c_i,d} = \frac{1}{n_{c_i}} \times \sum_{j=1}^{d} x_{c_i,j,d}, \tag{9}$$

where $1 < x_{center,c_i} < d$, n_ci represents the size of subpopulation in clan c_i, $x_{c_i,j,d}$ denotes the dimension d of the solution $x_{c_i,j}$.

The separating operator can be defined as:

$$x_{worst,ci} = x_{min} + (x_{max} - x_{min} + 1) \times rand, \tag{10}$$

where x_{max} and x_{min} denote upper and lower bound of the search space, respectively. $x_{worst,ci}$ is the solution in clan ci with worst performance, and $rand \in [0,1]$ is a random number.

By performing experimental tests with the basic EHO algorithm on standard unconstrained and constrained benchmarks, we have observed that the performance of the algorithm could be improved. One of the major drawbacks of the original EHO implementation is inadequate trade-of between the exploitation and exploration. Balance between these two processes are of the most importance for robustness and solution's quality of any swarm-based metaheuristics approach, and extensive studies have been recently performed in this domain [25–27].

In the original EHO implementation, the balance between intensification and diversification is shifted towards intensification. The process of diversification, that is performed by utilizing separating operator (Eq. (10)), is not enough, and in most algorithm's execution, the search process converges to suboptimal regions of the search space. To overcome this deficiency, in our hybridized EHO approach, we incorporated crossover operator from the GA. As already noted above, in this way we developed genetic EHO (GEHO) metaheuristics.

To enhance the exploration power, in each iteration of GEHO execution, two worst solutions are being replaced. The first worst solution is replaced with the new random solution obtained by utilizing separating operator, as in the original EHO approach. The second worst solution is replaced with the offspring solution from GA.

Algorithm 1. Pseudo-code for the proposed GEHO algorithm

Initialization. Create solutions in population; separate population into n number of clans; calculate objective function for every individual; set iteration counter t to 1, and limit iteration number to $MaxIter$.
while $t < MaxIter$ **do**
  Perform sort by fitness of all solutions
  for all ci clans **do**
    for all solution j in the c_i **do**
      Upgrade $x_{c_i,j}$ and create $x_{new,c_i,j}$ using Eq. 7
      Designate better individual between $x_{c_i,j}$ and $x_{new,c_i,j}$
      Upgrade x_{best,c_i} and create $x_{new,c_i,j}$ using Eq. 8
      Choose and keep better solution between x_{best,c_i} and $x_{new,c_i,j}$
    end for
  end for
  Sort population according to fitness
  for all c_i clans in the population **do**
    Replace first worst solution in clan c_i using Eq. 10
    Replace second worst solution by applying GA's uniform crossover operator
  end for
  Analyze and calculate fitness of all solutions
end while
return the best solution in the whole population

For the purpose of generating offspring solution, we introduced another control parameter (break point - bp), that controls the process of generating offspring solution. Before the bp number of iterations, the offspring solution is created by combining worst solution and pseudo-random solution from the population by utilizing the GA's uniform crossover operator with crossover probability (cp). However, in later iterations of algorithm's execution (after the bp number of iterations), with the basic expectation that the region with the optimal solution has been found by the algorithm, offspring solutions is generating by combining best and pseudo-random solution from the population, In this way, we did not only improve the balance between exploitation and exploration, but we have also enhanced the exploitation in later iterations by performing precise search in the promising search space region. For more information about uniform crossover operator, please refer to [25].

4   Empirical Experiments and Analysis

As it was already stated in Sect. 1, for testing purposes we utilized CloudSim environment, and EHO and GEHO control parameters were set to the same values as in [3]. This way we wanted to make a comparative analysis with other approaches that were simulated under same experimental conditions and under same problem instance more realistic. In [3], cloudy gravitational search algorithm (Cloudy-GSA) was proposed for solving load scheduling problem in the cloud systems.

For more information of task scheduling model that was utilized in simulations, please refer to Sect. 2.

CloudSim proved to be robust and elastic simulator of cloud environment and such it has been widely used by the researches. In experimental simulations, we replaced CloudSim pre-built scheduling algorithms with the EHO and GEHO metaheuristics.

In all conducted simulations we employed 25 solutions for dynamically scheduling execution of 10 tasks on 8 VMs. Specific characteristics and features of VMs and cloudlets are determined statically (hard coded) in the CloudSim simulator. Characteristics such as bandwidth, MIPS, execution cost and transfer cost have been used for determination of the total computational time. Similar, as in the case of Cloudy-GSA [3], EHO and GEHO utilize all functionalities of the system, yielding the total time complexity of $O(n^2)$. Results of simulations are generated same as in [3] where iteration set of values was from 10 to 100 and from 100 to 1000. Both approaches, EHO and GEHO are started 30 times. The reporter results represent the average value of the total cost objective as well as the transfer time.

The CloudSim environment parameters and GEHO control parameters that are used in experiments are summarized in Table 1.

With the goal of measuring real improvements of GEHO over EHO metaheuristics, the basic EHO, that was validated with the same parameters as the GEHO, was included in comparative analysis. Comparative analysis was performed with Cloudy-GSA, Simulated Annealing (SA), Genetic Algorithm (GA), Tabu Search (TS), Min-Min (MM), first come first served (FCFS) and particle swarm optimization (PSO). Results for all methods included in comparisons are taken from the [3].

Table 1. Parameter values in conducted experiments

Parameter	Value
GEHO parameters	
Number of clans (n)	5
Population size	25
Scale factor α	0.5
Scale factor *beta*	0.1
Total iterations ($MaxIter$)	10–100 and 100–1000
Breakpoint (bp)	$MaxIter*0.8$
Crossover probability (cp)	0.5
CloudSim configuration	
(x,y) coordinates	0–7
Number of VMs	8
Cloudlet number	10
RAM size	2048
Bandwidth of network	10000
Storage capacity	10000
Number of datacenters	2
Number of VMs in each data center	4

In conducted research, we executed two different experiments. First, we executed our metaheuristics with the goal of minimizing the transfer time objective. Later, as an objective function we took the total cost objective. For each experiment, the minimal and maximal value, mean and standard deviation in 30 independent runs of the proposed GEHO algorithm were reported. Comparison of the transfer times are presented in Table 2 while the Table 3 reports total cost objectives. The best results are printed in bold (lower values are better).

According to the comparison results, we conclude that in average, the GEHO metaheuristic has better performances in comparison with the other state-of-the-art methods. As an example, the Cloudy-GSA approach achieved better values than the GEHO only in the case of the mean for the transfer time and minimum of the total cost objectives. Also, the PSO performed better than the GEHO only in hte case of maximum indicator for total cost objective minimization. Also, from the presented tables, it is evident that GEHO obtains better performance than the original EHO in both tests and for the all performance metrics.

Visualization of methods comparison for the minimum indicator of transfer time and total costs objectives is provided in Fig. 1.

Table 2. Transfer time objective results and comparative analysis

Indicator	TS	SA	GA	FCFS	MM	PSO	Cloudy-GSA	EHO	GEHO
Mean	62768.22	63255.019	57502.225	59000	58000	54170.38	**50995.59**	53608.77	51513.34
Std.	0	7068.0319	2926.831	0	0	2505.163	3624.817	2850.107	2512.052
Maximum	62768.22	72822.8	62075.51	59000	58000	57759.06	58166.84	59613.22	**57619.91**
Minimum	62768.22	52661.52	51562.33	59000	58000	48525.77	43632.41	52302.53	**43117.51**

Table 3. Total cost objective results and comparative analysis

Indicator	TS	SA	GA	FCFS	MM	PSO	Cloudy-GSA	EHO	GEHO
Mean	160665.4	160423.0	154617.62	159662.3	156066.6	150759.4	148137.5	157320.5	**147905.1**
Std.	2.99E-11	7494.170	4192.3283	2.99E-11	2.99E-11	2290.429	2764.070	2932.115	2117.389
Maximum	160665.4	176870.7	163806.78	159662.3	156066.6	**154132.5**	155176.8	163250.1	154602.3
Minimum	160665.4	149520.3	147935.8	159662.3	156066.6	146814.2	**142157.8**	149295.6	142411.5

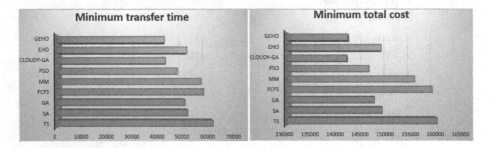

Fig. 1. 3D bar chart for minimum indicator of the transfer time objective (left) and minimum indicator of the total cost objective (right)

5   Conclusion and Future Work

In this paper we presented implementation of the elephant herding optimization (EHO) metaheuristics hybridized with the uniform crossover operator from the genetic algorithm for tackling load balancing challenge in cloud computing environment. Our approach, genetic EHO (GEHO), enhances the exploration-exploitation balance of the original EHO implementation.

Proposed approach was tested in the CloudSim framework, and two objectives were taken into account: transfer time and total cost. Comparative analysis was performed with other state-of-the-art metaheuristics that were tested in the same environment and under same experimental conditions. Obtained testing results proved that the GEHO is able to successfully tackle load balancing in cloud environment. In the future research, we plan to adapt and to test the proposed approach on other challenges from the cloud computing domain.

Acknowledgment. This paper was supported by Ministry of Education, Science and Technological Development of Republic of Serbia, Grant No. III-44006.

References

1. Rankothge, W., Ma, J., Le, F., Russo, A., Lobo, J.: Towards making network function virtualization a cloud computing service. In: IFIP/IEEE International Symposium on Integrated Network Management (IM), pp. 89–97. IEEE (2015)
2. Kumar, M., Sharma, S.: PSO-COGENT: cost and energy efficient scheduling in cloud environment with deadline constraint. Sustain. Comput.: Inform. Syst. **19**, 147–164 (2018)
3. Chaudhary, D., Kumar, B.: Cloudy GSA for load scheduling in cloud computing. Appl. Soft Comput. **71**, 861–871 (2018)
4. Buyya, R., Pandey, S., Vecchiola, C.: Cloudbus toolkit for market-oriented cloud computing. In: Jaatun, M.G., Zhao, G., Rong, C. (eds.) CloudCom 2009. LNCS, vol. 5931, pp. 24–44. Springer, Heidelberg (2009). https://doi.org/10.1007/978-3-642-10665-1_4
5. Kumar, M., Dubey, K., Sharma, S.: Elastic and flexible deadline constraint load balancing algorithm for cloud computing. Proc. Comput. Sci. **125**, 717–724 (2018). The 6th International Conference on Smart Computing and Communications
6. Mishra, S.K., Sahoo, B., Parida, P.P.: Load balancing in cloud computing: a big picture. J. King Saud Univ. - Comput. Inform. Sci. (2018)
7. Kalra, M., Singh, S.: A review of metaheuristic scheduling techniques in cloud computing. Egypt. Inform. J. **16**(3), 275–295 (2015)
8. Strumberger, I., Tuba, E., Bacanin, N., Beko, M., Tuba, M.: Hybridized moth search algorithm for constrained optimization problems. In: 2018 International Young Engineers Forum (YEF-ECE), pp. 1–5, May 2018
9. Strumberger, I., Tuba, E., Zivkovic, M., Bacanin, N., Beko, M., Tuba, M.: Dynamic search tree growth algorithm for global optimization. In: Camarinha-Matos, L.M., Almeida, R., Oliveira, J. (eds.) DoCEIS 2019. IAICT, vol. 553, pp. 143–153. Springer, Cham (2019). https://doi.org/10.1007/978-3-030-17771-3_12
10. Dolicanin, E., Fetahovic, I., Tuba, E., Capor-Hrosik, R., Tuba, M.: Unmanned combat aerial vehicle path planning by brain storm optimization algorithm. Stud. Inform. Control **27**(1), 15–24 (2018)
11. Strumberger, I., Tuba, E., Bacanin, N., Beko, M., Tuba, M.: Modified monarch butterfly optimization algorithm for RFID network planning. In: 6th International Conference on Multimedia Computing and Systems (ICMCS), pp. 1–6, May 2018
12. Tuba, M., Bacanin, N.: Artificial bee colony algorithm hybridized with firefly metaheuristic for cardinality constrained mean-variance portfolio problem. Appl. Math. Inform. Sci. **8**, 2831–2844 (2014)
13. Bacanin, N., Tuba, M.: Firefly algorithm for cardinality constrained mean-variance portfolio optimization problem with entropy diversity constraint. Sci. World J. Spec. Issue Comput. Intell. Metaheuristic Algorithms Appl. **2014**, 16 (2014). Article ID 721521
14. Strumberger, I., Bacanin, N., Tuba, M.: Enhanced firefly algorithm for constrained numerical optimization. In: Proceedings of the IEEE International Congress on Evolutionary Computation (CEC 2017), pp. 2120–2127, June 2017
15. Tuba, E., Mrkela, L., Tuba, M.: Support vector machine parameter tuning using firefly algorithm. In: 2016 26th International Conference Radioelektronika, pp. 413–418. IEEE (2016)
16. Tuba, E., Tuba, M., Simian, D.: Adjusted bat algorithm for tuning of support vector machine parameters. In: IEEE Congress on Evolutionary Computation (CEC), pp. 2225–2232. IEEE (2016)

17. Lal, A., Rama Krishna, C.: Critical path-based ant colony optimization for scientific workflow scheduling in cloud computing under deadline constraint. In: Perez, G.M., Tiwari, S., Trivedi, M.C., Mishra, K.K. (eds.) Ambient Communications and Computer Systems. AISC, vol. 696, pp. 447–461. Springer, Singapore (2018). https://doi.org/10.1007/978-981-10-7386-1_39

18. Sagnika, S., Bilgaiyan, S., Mishra, B.S.P.: Workflow scheduling in cloud computing environment using bat algorithm. In: Somani, A.K., Srivastava, S., Mundra, A., Rawat, S. (eds.) Proceedings of First International Conference on Smart System, Innovations and Computing. SIST, vol. 79, pp. 149–163. Springer, Singapore (2018). https://doi.org/10.1007/978-981-10-5828-8_15

19. Strumberger, I., Tuba, M., Bacanin, N., Tuba, E.: Cloudlet scheduling by hybridized monarch butterfly optimization algorithm. J. Sens. Actuat. Netw. **8**, 44 (2019)

20. Wang, G.-G., Deb, S., Coelho, L.D.S.: Elephant herding optimization. In: Proceedings of the 2015 3rd International Symposium on Computational and Business Intelligence (ISCBI), pp. 1–5, December 2015

21. Strumberger, I., Bacanin, N., Beko, M., Tomic, S., Tuba, M.: Static drone placement by elephant herding optimization algorithm. In: Proceedings of the 24th Telecommunications Forum (TELFOR), November 2017

22. Strumberger, I., Beko, M., Tuba, M., Minovic, M., Bacanin, N.: Elephant herding optimization algorithm for wireless sensor network localization problem. In: Camarinha-Matos, L.M., Adu-Kankam, K.O., Julashokri, M. (eds.) DoCEIS 2018. IAICT, vol. 521, pp. 175–184. Springer, Cham (2018). https://doi.org/10.1007/978-3-319-78574-5_17

23. Tuba, E., Alihodzic, A., Tuba, M.: Multilevel image thresholding using elephant herding optimization algorithm. In: Proceedings of 14th International Conference on the Engineering of Modern Electric Systems (EMES), pp. 240–243, June 2017

24. Wang, G.-G., Deb, S., Gao, X.-Z., Coelho, L.D.S.: A new metaheuristic optimisation algorithm motivated by elephant herding behaviour. Int. J. Bio-Inspired Comput. **8**, 394–409 (2017)

25. Bacanin, N., Tuba, M.: Artificial bee colony (ABC) algorithm for constrained optimization improved with genetic operators. Stud. Inform. Control **21**, 137–146 (2012)

26. Tuba, M., Bacanin, N.: Improved seeker optimization algorithm hybridized with firefly algorithm for constrained optimization problems. Neurocomputing **143**, 197–207 (2014)

27. Tuba, M., Bacanin, N., Beko, M.: Multiobjective RFID network planning by artificial bee colony algorithm with genetic operators. In: Tan, Y., Shi, Y., Buarque, F., Gelbukh, A., Das, S., Engelbrecht, A. (eds.) ICSI 2015. LNCS, vol. 9140, pp. 247–254. Springer, Cham (2015). https://doi.org/10.1007/978-3-319-20466-6_27

28. Strumberger, I., Tuba, E., Bacanin, N., Tuba, M.: Dynamic tree growth algorithm for load scheduling in cloud environments. In: IEEE Congress on Evolutionary Computation (CEC), pp. 65–72. IEEE (2019)

Author Index

Printed in the United States
By Bookmasters